LESSONS FROM THE OVERSEER

A Course of Study for Mankind

THE BOOK TREE
San Diego, California

ISBN 978-1-58509-137-9

Cover facial image
copyright by Louisanne

Cover layout by
Atulya Berube

Published by
The Book Tree
P O Box 16476
San Diego, CA 92176
www.thebooktree.com
We provide fascinating and educational products to help awaken the public to new ideas and
information that would not be available otherwise.
Call 1 (800) 700-8733 for our FREE BOOK TREE CATALOG.

I am the Overseer

I have returned

Dedication

To the blind masses of humanity living in darkness,
who have turned away from the light.

To those who are no longer asleep,
who can awaken them.

CONTENTS

LIST OF EXERCISES

Introduction

This book is meant to be a course in awakening. It can be done alone, with a friend or with a group in a class setting. It will be powerful for individuals on a personal level when done alone. Or it can impact things on a social scale in direct proportion to the size of the group, if a class in involved.

You, as mankind, have achieved many wondrous things with the powers of your mind. You have accomplished things that few could have ever imagined. At the same time, much of it has led to a deterioration of the planet, its resources, nature, quality of life, and your overall ability to control this destruction. As amazing and awe-inspiring as it is, the evolution of your intellect has out-paced the natural cycles of your priceless home, the Earth. You are highly intelligent, but that is not enough to recover. You lack the level of spiritual awareness that can, in turn, produce the moral character that will save you. That is the purpose of this course – to awaken you spiritually and help you build and express the moral attributes that will result from it.

Before determining your involvement in this course, however, the first thing any intelligent reader should ask is, "Who is the Overseer?" It is important to know the source of any information before deciding whether it is credible or not.

Allow me, therefore, to introduce myself. I am the Overseer and am very real. I have done my job for many centuries and am not a figment of someone's imagination. Nor am I your God – although some have mistaken me as such. I am not your creator, nor do I claim to function in any of the standard religious ways that you recognize in your views of God. I am below God, and act as an Overseer of His creation. I am a mere messenger to and from the highest God, and hold a more lucid view of the divine than your human family possesses. My purpose is to share with you insights into who you are, what you are doing on your earth, God Himself, the physical world and the truth behind reality as you know it.

God, in the entire scheme of things, is your Father. The best analogy to use in describing my relationship to you is that of a big brother. Humanity is much like a rebellious teenager. They think they know everything, but continue to get into trouble. For example, teenage males sometimes have trouble with the law and display too many violent tendencies, while the females sometimes make mistakes they must deal with for the rest of their lives, like running away, shoplifting or becoming expectant mothers before they are ready. Teenagers are often in trouble because they have not matured enough to make informed, intelligent decisions yet suddenly find themselves in the position to do so.

As a big brother, I have a larger perspective and have matured beyond your limited scope. However, I cannot make your decisions for you. You must learn and grow for yourselves. As a species you have finally begun to see what direction you need to go. Therefore, I can shine a beacon of light toward you so that you might follow it out of the darkness, reach a better path to travel, and learn your lessons more easily.

The most difficult part in explaining myself is that I am not a normal human being. Therefore, you may be reluctant to give me credibility. My words may strike you as New Age nonsense, or they may ring true. You will be unable to form a good opinion, however, until you read a bit further. That is all I ask. I do not ask for acceptance; that is not important to me. The overall response you give to this course will let me know if you are ready. If humanity is not ready to listen to the lessons, then it is simply not time yet and you are not ready.

An individual like yourself, however, may be ready. If so, you may reap enough wisdom to contribute things to your world that may make a positive difference toward the collective good. You will be capable of making positive changes in the world by acting on the lessons in this work, if you should find it valuable. Regardless of what you may think at the moment, I do have access to information that may prove valuable to members of your struggling race, and I am willing to share it here, within these pages.

I have been here for centuries, watching and guiding you, but never doing things for you. You must do those things for yourselves. I have *not* appeared in the physical for many of these centuries. I do not normally show up and talk to people. My guidance manifests as important ideas through various people, with some of those insights – but not all of them – appearing in your reality after hard work, tenacity and creative ingenuity by those who receive the information. With some of you, the inspiration comes through in a flash and appears in the mind instantaneously. I choose who is best to receive the inspiration rather than appear myself to share it, and calculate through means unknown to you how best to achieve the goals that serve humanity as a whole. You have never been exceedingly receptive, but I continue to try.

The time has come to present ideas to more people through a book of teachings, instead of targeting receptive individuals. The need for bigger changes, stemming from a crisis in humanity, requires it. It is time to present ideas to everyone, through written form rather than thought forms. That is the purpose of this book. It is meant to combat a crisis that is upon you. There is an urgency that is much larger today than in the past, with more action required by more people to overcome your problems. A physical book is more easily shared and can be a true catalyst for what you need.

If you do not believe in guidance or inspiration from outside sources, it would be best for you to put this book down and walk away now. Do not waste your precious time any further by challenging your rational thought patterns; it won't work. If, however, you are open to the appearance of sudden inspiration by reading a book, and if you are open to making a positive difference in the world as a result of that, then this book may be of value to you.

What may result with individuals will be a gradual transformation. Some of you will get further along than others, but that should not be a concern. By going through this course, you will end up being exactly where you need to be, so comparing yourself to others will not be necessary. Everything will be in sync when you reach the end, whether you have a complete "transformation" or not. You will, among other things, experience a powerful humility. This may sound like a contradiction in terms, but once you experience it, you will understand. Humility is one of the more valuable human attributes that is sorely lacking in your world. This is just one area of many that this course shall address.

This course is a guidebook for the human group that inhabits what they call the Earth. Some of them may be ready to listen to these words, which is good. This course is for them. They will become inspired and begin to make positive change. In a collective sense, their media authorities are not willing to present this course in any direct way to the masses, and promoters of this course would be ridiculed by the so-called "rational" masses. But the masses still operate from a level of "teenage maturity." The masses do not have the means to hear or act from this book to any effective degree, due to the level of ignorance and disorganization that surrounds them. This course is being given so that the few who benefit from it will pass these lessons on, so that at some time in the future it will have an effective impact. Awareness can be created out of the inspiration of those who understand this book. The Internet can be a helpful alternative to the closed-off, mainstream avenues as it provides a voice to those who will act and a means of contact for those willing to listen. Mankind must act soon. I have patience with the human race, but it runs short in the present climate – one that is beginning to teeter on the brink of destruction for mankind and its' supporting environment.

Some might ask, if things are that bad and mankind could benefit from outside help, why should I not appear and just give it to them? Providing you with such help is the point of this book, which can be done from afar. Physically coming down and solving humanity's problems would cause worldwide fear, distrust and even panic, due to your primitive superstitions and divisive religious mindsets. If these barriers were somehow circumvented, it would still not result in a lasting solution. Humanity must solve their problems for themselves and by themselves. That is their purpose, their mission, and I cannot interfere directly in that process. Doing it for you will not transform you. I can only offer indirect guidance, which can be used by inspired teachers or leaders with wisdom when humanity decides it is time to put such leaders in place or, more importantly, when teachers step forward and take matters of spreading such wisdom into their own hands. Becoming receptive to more holistic and compassionate ideas are of no use unless actions, rather than words, are used globally. These actions *must* take shape with the personal initiative of humankind in order for them to be effective.

We – meaning those who are like me – had our attention drawn to the Earth due to the immense level of noise that it emits throughout the universe. Even without your technology some noise was always there, due to the life your planet harbors. However, the noise has increased greatly with mankind, especially in recent years. The Earth is immensely polluted with noise-making apparatus, which now draws the attention of all advanced civilizations. In almost every case, they look sadly upon your world and will not make open contact with such a backward race. We cannot help you; you must help yourselves, despite the fact that you are failing as stewards of your planet. Approaching your planet is a sight to behold. It is a stunning, blue jewel shining in the sky, replete with life-giving moisture and unique with its beauty, tectonics, and myriads of life forms. But as one draws closer we behold the orbiting space junk, damaged ozone layer, impure atmosphere and, upon closer examination, your polluted waters on a planetary scale. The Earth itself is a living being, but has been taken over by an insane species that is literally choking the life out of her. The Earth has certain wisdom and associated power through the movement of nature, however, and will not, under any circumstances, allow herself to die. Mankind will be the ones to die, if it should ever come to that. And such a death has been the case with many other dominating Earth-bound species in the past, some known and others not known, by current Earth inhabitants. A time is drawing near whereby the Earth in her wisdom may be forced to move once again in an act of self-preservation. Worldwide cataclysmic events are immensely terrifying and messy, but not uncommon. They serve a purpose. When the Earth slowly cleans herself in the aftermath of such an event, it is the ultimate act of cleansing and purification. It takes many centuries, but the Earth knows exactly how to repair itself with new life and, if necessary, with an entirely new atmosphere. To avoid such a scenario, Earth's inhabitants now have a choice, and a chance, to provide that cleansing beforehand. Otherwise, humanity may be effectively removed. You can continue to be the source of the world's problems, for both yourselves and for the Earth and all of its inhabitants on land and sea, or you can begin to make important strides in reversing negative trends and providing a solution.

In ancient times the gods knew that a great flood was coming. Most of them were willing to let humanity be wiped out, but one god decided to warn humankind and, in effect, save them by sparing a few. Those few were then able to repopulate. The story was preserved in a series of ancient Sumerian tablets now known as *The Epic of Gilgamesh*. The hero's name in this tale was Utnapishtim, who became known, centuries later, as Noah when the story was passed down.

A similar situation has almost been reached today. I have chosen again, as the Overseer, to extend this information to mankind – this time leaving it up to you, rather than the gods, as to what will happen next regarding your potential survival. Mankind has matured to a point where it can make its own decisions rather than having them made by gods who happen to like your planet and may not have your better interests in mind. Such decisions made by mankind may not be good ones, but at least you have the right to make them for yourselves and, for now, that right is being respected. This book is to assist you in making those decisions in hopes that your time and safety on your world may be extended – and that serious dangers may be averted.

If you are taking these lessons as part of a class, it is suggested that you read one chapter each time, do the exercise associated with it consistently before your next meeting, report on your perceptions, ideas and results in a discussion, assess what it means and where you see things going, and continue doing so each time.

This course contains only thirteen lessons, as listed on page eight, so you can go to them whenever you need to. It is better to master a small number of powerful lessons and do them consistently, over and over again, than to experience a long, continuous list of lessons, just to keep trying something new. This book is not meant to be a novelty. It is a serious and powerful course that can completely transform your life and potentially, the planet.

Some of these lessons are deceptively simple and when you encounter them, especially near the beginning, you may wonder what good they really are. The early lessons, however, set the stage for those that come later – and are therefore equally important. This is a *process*, which must be carried out sequentially in order to provide you with the results that you seek. So embrace and honor each lesson; do not skip them, but experience them fully.

This course can make a great impact, depending on how you use it. To get the most from it, it is advised that you use a calendar to "map out" your activities in relation to it. When an exercise advises you to perform it at certain intervals, mark them on the calendar so you do not forget. Stay with the program, and you will experience the greatest results. Once you have completed this program, a certain "cycle" of the exercises will be there on your calendar. This will allow you to go back and repeat this cycle once or twice a year, or continuously, if desired, using page eight to find each exercise when needed. You will also end up refining, changing and improving your work as you go. Without the recommended repetitions, or without doing the exercises regularly from a comfortable schedule that you create, this will become just another self-help book that you had fun reading and experienced a moderate impact with. Most of your self-help books and courses do this, and you almost always revert back to your normal, ingrained habits soon after you finish the program. Deeper changes must be achieved before you can simply pick up a book or go through a course that will completely alter your personal development or social structure.

This book can move you toward those deeper changes. It does not claim to contain magical answers. Only you contain those answers. This book will open you up so you can reach deep inside and bring out the answers that you need. These lessons are merely a catalyst and, unlike your countless courses and other books that claim to be "the answer," this book claims nothing. You must dig deep, find out what changes are best made for you, personally, socially, economically, and *make them* based on what you read. Your actions will follow an inner "trigger," which I can provide, but you must pull that trigger. Your actions are separate from this book and come from you alone. So read the book. Take its lessons to heart. And act. Any and all credit for the amazing things you accomplish from reading it is yours and yours alone. This book is nothing but stagnant words on a clump of inert paper or a computerized screen. It has no power. Only you do.

PART ONE

PERSPECTIVES

Lesson One

Perspective of the Overseer

The first lesson requires you to know your teacher and the perspective he has. What qualifies me as your teacher? How do I perceive you and your problems?

Being the Overseer, I am able to glide over long distances from above and examine from the air exactly how humankind has laid out its' cities and social structures. It is easy to see the grid, or overall pattern, being used. The physical grid is a reflection of the psychological status of mankind. The outer world is a reflection of the inner mind of man. From on high, your various cities look like electronic circuit boards in the way they are laid out. The human mind is laid out in a similar internal "pattern," having been programmed to perform certain functions. Predictable patterns then emerge in the outer world.

When city lights are lit up on the physical grid, it reflects quite closely the same general circuitry of the human mind. One reflects the other. Therefore I, as the Overseer, can clearly see what is happening or is about to happen with humanity by merely glancing down. Your level of consciousness is displayed through your visible technology and architecture. Your ideas of territory, protection, progress, compassion, aggression, morals, ideals, self-preservation and every other important component of your thinking is clearly apparent to me. I can see exactly where you are going and why. Just as a helicopter can see far enough ahead on a train track and radio a warning to the conductor, I too can do the same for mankind. Just as you can look down on a mouse inside a maze, I can see exactly what options are there for you on your path and, despite what you might be thinking or are concerned with, I can see what options do *not* exist and are *not* there for you.

The human ego often insists there are answers that can be willed into existence when there is a dead end in the maze and no such answers exist. The good news is that some of these answers regarding your survival still exist; while the bad news is that the window has closed on some of the better answers, and is closing quickly on the ability to access the remaining ones that you need. I cannot give you these remaining options in a way that will cause a unified and positive response; such a unified response will only come, unfortunately, in the form of catastrophe. I am reluctant to create catastrophe because I respect all life, although I cannot rule out such actions coming from others from where I dwell. Nor would I move to prevent a *natural* catastrophe that may be coming because it is part of nature's cycle, or would be a protective response from the Earth herself. I honor nature as the instrument of God. Such events must be respected, as they come from an immense and ancient intelligence that few humans understand. The possibility of a protective response from the Earth in the form of a catastrophe is quite obvious from the disharmony that mankind has created with nature.

Your noise permeates the entire universe – at least to the extent that it has traveled so far. When approaching your planet, the noise level increases dramatically. This is like a farmer who, with a higher form of consciousness than chickens, opens the door to a large henhouse to find intolerable and meaningless squawking. We find the same thing on a smaller scale at a sports stadium with a screaming crowd. The crowd is focusing on the only thing that matters to them – their team "winning." People are suffering and starving throughout your world on a massive scale, but those who can actually do something about it are *screaming for an illusion* instead of focusing on reality. It doesn't matter that one "team" with different colored uniforms scores a few points more than another on any given day. You are all equal; you are all God's children who should be helping each other and working together as *one* "team" – the team of humanity – with compassion and mutual respect for each other, rather than pitting yourselves against each other to scream for an illusion. Based on your incessant need for ego gratification, to create a false sense of superiority, you have created endless variations of these competitive illusions – not only through sporting events, but through competing religions and the accumulation of wealth, property and material possessions that reflect "status." The loud and trivial *noise* that you create is the result, and fuels the pursuit of these things.

Everything is a vibration. All vibrations must have a frequency to operate on. Frequencies make noise. All communications that you make, no matter what they are, make noise. The devices used, like cell phones, computers, fax machines, televisions and more, make noise just by being turned on. They do not have to be communicating information to be making noise. All of these devices are emitting frequencies just by being operational. When they are transmitting data even more noise is there – and the amount of noise and frequencies that bombard each other throughout your atmosphere is causing chaos throughout nature. Before these electronic devices existed, nature had her own frequencies. For example, whales in the ocean try to follow a form of these natural frequencies and once migrated safely. But with high tech military frequencies resonating through the ocean depths, pods of whales get confused and are commonly known to beach themselves and die because their natural navigational abilities are disrupted. Water amplifies these frequencies to a great degree. So your unnatural frequencies have penetrated the water and the air of your planet. There is no escape. If you could see the energies bouncing all around you in your atmosphere you would be appalled at what you have created. You have no idea what the effects of all this really is, and I am here to tell you that it is bad. Pollution does not have to be visible

Mankind was once part of nature, living within its rhythms like all other beings. This is the way nature intended. However, humanity became too "clever" for its own good and outsmarted itself. Every serious problem you have on your planet, no matter what it is, leads in one way or another to your disharmony with nature, or your outright exploitation of it.

There is only a finite amount of exploitable resources available on the planet that can be mined, pumped, processed or sold for a profit. Those who control those resources control the planet and, to a certain extent, control all others – who must answer to them for these resources. Therefore, what drives the economies of the planet also depletes its resources. Due to the demand, these resources are consumed at an alarming rate. A good example is found in the rain forests of South

America. They provide much of the Earth's oxygen, but are being cut down so fast that the resulting lack of trees is showing a decrease in the levels of oxygen in your atmosphere. You are suffocating your own "mother" – the Earth – and, in turn, yourselves. Your concern, however, is not with having enough air to breath or in repairing the damage that you have done, but in trying to figure out what other natural resource you can harvest and sell next. You will have depleted your planet of every valuable resource very soon unless you can develop alternatives. This can be done in some cases. The one resource that cannot be replaced with an alternative, however, is *water*. Water is the very essence of your life and there is no replacement for water.

Your biggest mistake, whether you know it or not, has been your gross misuse of water. Most of the world's rivers as I see them from above are clogged with garbage and feces and are unfit to drink from. Only a small percentage of your fresh water, apart from your polar ice, is still clean in the way it should be. The oceans are just as bad. The middle of the Pacific Ocean has a huge swirling dump of floating garbage and debris over one hundred miles wide at the precise point where the eddies and currents of migration meet, creating this swirl of inescapable garbage that has destroyed much of the sea life in this area. In other areas coral reefs, which are essential to sea life, have dwindled greatly due to mankind's mistreatment of water. The lack of such reefs has destroyed the economies of small fishing villages, tourist locations, and others that were dependent upon the presence of these reefs for centuries.

In major cities throughout the world water treatment plants filter and recycle wastewater and then make it available again to your people. However, the filtration process is not advanced enough to eliminate residual pharmaceuticals and medications used by other people for things like fertility and other unwanted effects that you would only want to happen to your body if a doctor prescribed them. Water is becoming your biggest curse, in addition to your most cherished resource. It is the key to life and you have taken it for granted. All of the ways you have polluted your water sources show that you have placed material gain over the importance of your own health and well-being. You are not so much suicidal. You are simply neglectful and ignorant. You are focused more on personal material gain through exploitation rather than accessing and using a powerful collective consciousness to help yourselves as a whole. That is how I view humanity.

Exercise

Higher Perspective

You must do your best to develop my perspective. Distance yourself from your normal sources of information and develop new ones from outside your comfort zones. Your goal is to view yourself and your world as an outsider from afar would do.

First, abandon all your normal information sources. This means everything. All local TV and radio stations are off limits whether you used them normally or not. You should watch or listen *only* to foreign news sources for at least 30 minutes a day for each source, preferably accessing two separate ones for a total of one hour. If you do not have cable television, short-wave radio is an excellent alternative. If you have both sources, use them both. The Internet may also have some interesting visual news feeds. Pay special attention to 1) news that relates to your country and its influence on the world and 2) to news about humankind in general and its influence on the world. Do not cut yourself short; it is important to examine these news reports deeply and carefully, with new eyes and ears.

After at least three days of doing this you should have a slightly expanded view of where you are fitting in the world. It may take longer; do not rush. Once you are secure with this knowledge, it is time to move on to the last part of this exercise. Do not begin doing this until the first part is complete.

You have spent your life being taught that you are from this place or that place, and "belong" in certain locations. That is, until you move – then the process starts all over again. You identify yourselves with certain historical landmarks, documents and sayings in order to cultivate pride, give you the desire to function within its laws, pay its taxes and, if need be, fight to the death for it. But in reality, you are not from "here" at all. A bigger picture exists.

Before you were born you were from somewhere else, and when you die you will return there. When an old and wise person is expecting to pass away soon, they are often heard to say that they are "going home." Those who have lived long and productive lives have this realization more than others.

Become relaxed. Close your eyes and repeat softly from ten to twenty times, "I am not from here, my home is somewhere else."

Although where you are from is in a far safer realm, being spiritual in nature, you need also understand there are other physical realms than here that support intelligent life. Some of this life is far advanced to yours and, should I dare say, *alien*. With this in mind, you are to examine all television, radio and computerized programming. Block out all personal associations that you have developed and experience these means of communicating as if you are an alien being who has just discovered them. You are intelligent enough to decipher what is being said, but you are seeing it for the very first time. You are exploring this world from afar (as some are actually doing) and, with your advanced form of logic and intelligence, are assessing what is really happening on this world. Pay attention to the levels of cohesion and division that you encounter, and what solutions would exist from your higher perspective.

You will find great beauty, joy and compassion in your study; coupled with greed, ignorance and cold-hearted manipulation. Spend another three days studying this new world, and from this "clean, uncluttered slate," reveal this experiment and discuss your observations and findings with your friends or family. As a result, you will learn even more from them because *your experiment is continuing*. It is not over. During your discussions with them, pretend that you are still this alien being, now in human form as your "self," and study their responses. Finally, if you are reading this book as part of a group rather than individually, return and discuss *everything* with your fellow "alien being" classmates.

Once you have completed this entire exercise, you will have a better understanding of how I perceive you. Step one on your journey will be complete. The next Lesson will cover how you perceive yourselves.

Lesson Two

Perspective of Humanity

This lesson will clarify how you view yourselves and what can be done to change this limited view for the better.

As I glide above the planet, over your homes and factories, your streams and rivers, the jammed traffic on your roads, the crowded, unsanitary conditions of your cities, and over the diminished but pristine beauty of nature still found in those few far corners of your world, it has become clear that humankind has spread itself like a cancer upon the living Earth. When this disease strikes a human, it is the person who must act in an attempt to be rid of the problem and destroy the cancer. The Earth is no different. She knows, however, that she really is your "mother" and is reluctant to destroy her children unless pushed to the inescapable point of self-preservation. It is uncomfortable for nature to perform this duty, much like having to cut off a cancerous limb in order to save the organism.

In rare occasions cancer will disappear by itself in humans and when this happens, it is called a "miracle." You are capable of miracles. You all have an indwelling human spirit that is capable of great things if you focus and apply yourselves. The vast majority of you do not know how to harness this energy. You instead look feebly to others for your answers, not knowing that you have them within yourselves. Each and every individual who reads this must not depend on others, but act in ways that reflect the fact that your very survival – your personal survival – is dependent upon your actions. There is no such thing as being "lost in the crowd" or simply blending in to avoid responsibility. You are responsible and each of you does make a difference. You have a more vital impact on your surroundings than you know. The idea of being just one individual out of many, who could never make a difference, makes it easier for you to do nothing. That idea is a dangerous illusion. Your proper perspective of meaningful, responsible action has been numbed and masked over by interests that are manipulating you when in fact you should be fully conscious of your health, well being and survival. Blind comfort is no substitute for *conscious awareness* because eventually, your blind comfort will destroy you as a race. You are already beginning to feel the uncomfortable consequences of your disassociative malaise. The discomfort that this blind mindset causes is a prerequisite for your destruction. It is now upon you and has already begun. It is time to wake up. It is time to change your way of thinking. I do not know if you can do this. I can only urge you to try.

I can see many things from above, but one thing I cannot see is the future. I can see the past with great clarity, however, so I can reveal to you things about the past that may help shape your future in a positive light. You must go back and study the Native Americans, as they were before the advent of what you call progress. They lived off the land and nothing, absolutely nothing, was wasted. Today you live in the most wasteful society the world has ever known.

One styrofoam cup is a good representation of your culture. The planning, manufacturing process and shipping of this cup is time consuming and expensive. Dies were cast to create its shape, factories were built to produce it, the machines use energy to make it, the act of producing it in an assembly-like process takes time, and then it is stacked with others and boxed. People are needed to package them, trucks are needed to ship them. When it finally gets to the store they must be stocked on the shelves. When the cup finally gets used, its purpose is fulfilled in less than a minute, sometimes in just a second or two. A long-distance racer, during a race, will grab one and splash his face with water in less than one second before it is thrown to the ground, its purpose fulfilled. Compare the entire process needed to make it to the time it was used. Then it goes to a landfill and will be there for hundreds of years before it will break down and be gone. Millions of these cups alone pollute your Earth needlessly. There are other alternatives. Plastic water bottles are even worse. Because you act in this way, you are surrounded by negative consequences that are surfacing all around you. You have been trained to think as "consumers," and care for nothing except your own personal convenience. Your selfishness has eroded the very fabric of your existence and you are now being forced to think differently and hopefully, act differently. Your survival is now staring you in the face.

For thousands of years the Native Americans lived in harmony with the land. In just two to three hundred years, however, you have almost destroyed yourselves and the environment with your suicidal swan dive of existence. This lifestyle has no place in a world that naturally, on its own and aside from humanity, operates through a system of harmony and balance. Nature knows how to take care of herself. She and the Earth have been around far longer than you have, meaning mankind itself. If you credit yourselves with ten thousand years of Western civilization, which is a very generous credit, that is only about one half of one millionth of the age of the Earth.

Do not worry about "saving the planet." Worry about saving yourselves because nature, or the Earth, will do away with you rather quickly if you squander your existence here and fail to take care of her. That time may be coming soon, unless you change course. After this happens, she will cleanse herself of you so thoroughly that it will be difficult for future life to determine that you were even here, because she will take her time recovering from her response to your onslaught. Having "all the time in the world" is clearly not a problem for her. Yours, however, is running out.

Here is an assignment for you. Watch the film Koyaanisqatsi, which was made in your year 1982. Rent it or find it in any way you can. The executive producer was named Francis Ford Coppola. This was before he became famous with more mainstream movies. The film has no words, but tells a far more important story than any speaking film could ever convey. You must learn to be more observant of the things around you. All of you. My eyes see everything. There is a scene in this film with a newscaster who has TV monitors behind her. If you slow down this part and watch the background monitors, frame by frame, you will see a number of different subliminal images that cannot be detected at full speed with the naked eye. Study them. I will leave the interpretation up to you. This entire film relates largely to the Hopi Indians, their reverence for the Earth, and their prophecies. Some of their detailed prophecies have already come to pass.

It is true that Native American tribes were not perfect. They were warlike, as you are today, and life was harsh due to the increased exposure to the elements that they had to endure. But being closer to the elements brought them closer to nature and they knew better who they really were. That is the important part. Every day required actions that their survival depended upon, so if they failed in these actions they would suffer immediate consequences. Life, therefore, had more meaning because their daily actions had more of a direct impact on the comfort and well being.

Today, on the other hand, you have grown fat and lazy. Your daily actions don't mean as much because you lack *the direct responsibility* for them, and the immediate consequences that result. Therefore, you do not feel as alive and invigorated as the Native Americans did. You need to wake up and be alive again.

Your minds are focused on pursuits that take life rather than give it, because it makes money. For example, your government contractors have created an industry of warfare-based weapons, so warfare must be waged to keep the economy strong. Or that millions of species have gone extinct at the expense of corporate giants, or that people do not need all of chemical pharmaceuticals and food additives that your culture feeds you, when more healthy, natural alternatives exist. Your modern cultures need you to be unhealthy so that large companies, masquerading as nature, can provide your "answers" in place of her, and continue to make money. The economies you have created require it. But natural cures for your ailments, without the chemicals, have worked in places like China, Siberia and the Amazon jungle for thousands upon thousands of years. Your so-called backward, primitive shaman hasn't a nickel to his name, but can run medical circles around many corporate-backed doctors with millions of dollars. So you fear these native people. You insult them by calling them primitive heathens and "witch doctors" because they don't have a medical degree from Harvard. Their natural cures, if you even know about them, will never be promoted by your societies or be given credibility, despite their proven results through centuries of trial and error refinement.

You have trapped yourselves in a deadly plastic environment, fake in almost every way. In fact, your landfills are full of plastics that do not break down or dissolve naturally for hundreds, or even thousands, of years. Your cities do not know what to do with all of the garbage that you create at break-neck speed – consuming, consuming, and consuming. You are trained to be nothing but consumers, spending your money on countless things that you do not really need, but are *told* that you need. You listen to "authorities" and react to their commands because you have never been taught to think for yourselves. Native Americans had to think for themselves on a daily basis, because their survival depended upon it. They were adept at living with nature, but were ultimately no match for the European interlopers who failed to understand it. The Spanish conquistadors and white Europeans completely decimated native tribes after terrible treatment for hundreds of years – primarily because of a complete disregard for the workings of nature and the misunderstanding of humanity's place within it.

The idea of "owning" land was incomprehensible to Indians. In their eyes, only the creator had title to it, who allowed all to use it. You cannot own the sky or the air any more than you could own land. The territorial, warlike attitude of the Indians was still present, so they were not perfect. They were flawed with violent attitudes at times, but it was not about land ownership. It was about the *usage* of the land during opportune times. It was about positioning themselves in the best natural areas for hunting and fishing – areas that would allow them to live in harmony with their environment and function as a part of the natural process. You cannot own land because you cannot own nature. They understood this.

If you can relearn this, the world will truly be your home. For now, your home is only one tiny area with walls that you have a deed to or pay rent or a mortgage on. But the entire world should be your home, shared among all with equal respect. You can comprehend this, with difficulty, but it is far from being realized.

Your perspective is very limited. You believe that the whole world is at your command, being the most "intelligent " species and therefore, with the most control over the environment. However, in reality, you are delusional and it is the other way around. The world has command over you. Nature still has command over you. But you constantly fight it. Humankind was once part of nature, but due to a growth spurt in the brain, new ideas took hold that created your cities and all of the other huge developments that you interpret as "progress." You have created separation and ownership with your land, making borders that others should never cross without your permission. These states, counties and cities sought, and still seek, a way of life that separates you from nature. But without nature you destroy yourselves. You truck in food and pipe in water to your cities while losing direct connection to these life-giving sources – the land. The land gave you your very lives, it is the root to your existence. Your separation from it is a numbing illusion. Your connection to nature is deep within you, hidden in the past and within your genes, among the countless generations of those who lived in harmony off the land. It is in you genetically and within you subconsciously, and is as necessary and vital as the air you breathe. Despite trying to collectively choke off this connection, it is still there, clinging to you as *a hidden strand of life* because without it, you would die. And you probably will, if you keep ignoring it without nurturing it.

By separating yourselves from nature you have created the illusion that you are separate from the events that occur within it. You attempt to control it with the mistaken belief that the consequences of your actions have no direct bearing on your lives. The fact is that no man is fit to control nature unless he feels himself to be part of it.

You have failed to see life holistically. Only now is that beginning to change. You currently spend the majority of your time indoors, away from nature, staring into a box called television that delivers images of the world to you. Or your "fix" comes from a computer screen via the Internet. These images make you believe that you no longer need to go out and experience life for yourselves, because you are lazy and would much prefer that you have vicarious experiences delivered directly into your homes on a silver platter. Large corporations that would otherwise have to work hard to convince you to buy their products love this, because they know one thing. Lazy people are usually *stupid* people, so these large companies design what they call commercials. These advertisements insult the intelligence of smarter individuals, but are designed to convince average humans to buy their products when you do in fact venture away long enough from your mesmerizing boxes.

Your other box, a computer, allows you to buy almost anything from your square shelter without ever having to leave it. The only time many of you leave your homes is out of necessity. Working at a job is the main necessity that requires you to climb into a motorized box on wheels and drive away from your home so you can sit inside a cubicle, inside a larger square room, inside a square building, often looking at another computer-box, or packaging items into square boxes to be sold. The list goes on and on as to how compartmentalized your world is. When you live like this, you think in the same limited way. That is the point. Your minds are now boxed up and dead, apart from the expanse and beauty of nature, which is *alive*.

You put yourselves into psychological boxes based on your environment and are trained to respond over and over again in ways that *stunt your growth*. Your square house with its square rooms sits on a block, within a squared off, boundaried city, county, state and country. A personal change in this structured environment would release the boundaries and free your minds. But you fail to see it, so you fail to act. This failure will lead to self-destruction – unless something is done to wake you up.

Most of you, despite the obvious trouble and discomfort you are experiencing with yourself or in society, do not *want* to wake up. You are masking over the problems in today's world with technology. It is your crutch. It has happened to such a degree that you are not thinking in healthy ways. You have fooled yourselves into believing that technology is healthy, while it is the exact opposite! You are ingrained in your technology too deeply. You believe that you can solve the world's problems with technology, but these times call for a return to more simple and basic solutions that require an *abandonment* of your technology – not a headlong race into more, and more and more. The main message you are being fed is, "Buy the technology." No matter what it is, or how bad it is, you are being brainwashed that it is good.

For example, you now have genetically modified food. Seventy to eighty per cent of processed foods sold in supermarkets have genetically modified products in them. They were originally created in laboratories to resist the natural diseases of crops and to solve world hunger, according to the companies who own these new "designer" foods. Studies show no decrease in world hunger, but an increase in the profits of the companies that hold the patents to these unnatural forms of food. Many of these genetic variations were tested and caused serious health problems in either animals or humans, so were abandoned. Other variations have not been tested properly and in some countries, including the USA, they do not have to be identified through labeling as being present in your food. You are tampering with genetics, patenting bizarre new strains life, rushing them to market without adequate testing and then *eating it*! What are you, crazy? (You do not need to answer that, because the answer is clear.)

Only nature and/or God should create life. Not you narrow-minded, egotistical humans who blindly swing into action whenever there is a profit to made, who know nothing about life's true origins and have continued to ruin the environment. The simple answer is to care for one another, rather than exploiting each other for the sake of profits, and go back to nature – not only in this instance, but in countless others. That was only one example out of thousands. Some of you claim that you are no longer savages; that you now live in a "civilized" world and all of that is behind you. It is just as savage, if not more so, to exploit others and watch them suffer while you reap profits at their expense. This is knowingly done, while armies of lawyers are hired to protect the wrongdoers. Your battles are being fought today with guile and money rather than primitive weapons and because of this, the damage is actually more brutal and severe. It is more difficult to stop the bleeding when the victims cannot tell where the arrows are coming from. Instead of loincloths and clubs, you wear ties and carry briefcases. You cannot track such a perpetrator to a physical cave or castle, because the modern corporate manipulators hide behind computer networks, security cameras and boardrooms encased within glass towers. Money buys their security. The horrendous deeds they initiate are performed by others in ways that hide their involvement. And you proudly call it progress.

You cannot go back to your more primitive ways, however. Too much has happened. You cannot return to nature, completely abandon all that you have accomplished and live like savages again. That is not my point and that is not the answer. This progress that you have experienced also has its good points. It is part of a learning process for mankind. As the importance of nature becomes clear to you, after your long separation, the time will soon be upon you *do something about it*. Nature, like a long-lost lover or family member, will not take to being spurned for any length of time without consequences. You cannot treat someone poorly for so long and expect them to come running back into your arms after you have damaged them. Nature has intelligence. It will not do this and can take care of itself, if need be. It has the wisdom to know how to heal itself if the problem (humanity), intelligent as it claims to be, cannot salvage the situation itself.

When you were part of nature many centuries ago, things were in balance. You lived as tribes in your natural surroundings, moving through nature on a journey. When things became depleted you would move on and find new sources of food. At times, with food being constant and plentiful, tribes would stay for very long periods. Later you created agrarian societies, settling into one specific location with the advent of farming. There was no need to move when a society developed ways to find or create everything it required without having to pull up their stakes and move on to an uncertain future. As more and more people became lazy and dependent on this lifestyle, knowing they would not have to move, the available food could no longer support their numbers. The farmers needed more land to sustain the people, who were too lazy to move out, so it was decided that the farmers would be "exiled" to greener pastures, away from the pining hoards, to create as much food that would be needed. The farmers did not mind because they would be paid for this food as an act of commerce. Instead of the entire community sharing in the venture of farming, which small agrarian communities and communes could do, larger groups that qualified as towns began doing more than just farming.

The birth of towns made life interesting for you, but it was the first major step away from nature. You have been learning how to cope with your estrangement ever since. Nature has given you this opportunity to learn, but at her expense. She has suffered and continues to do so at the hands of humanity. Since humanity is actually part of nature, she has allowed her errant children to wander, hoping you will learn enough on this journey to understand *balance*. Then you will come back to her – at least it is hoped.

Your perspective must change. If that happens, however, you must act on it. If you achieve a new perspective, your actions must reflect it in ways that will reverse some gigantic trends.

Exercise

Inner Perspective

In order to understand yourself, you must cut straight through the mind. The mind is not only a filter, but is a barrier to the Self. You must therefore bypass all of your thinking and analyzing and studying, and go directly to your true "self" and the reason you are here. The only way most of you have ever understood *anything* has been with your minds, so you are probably asking how you can understand yourself beyond the function of the mind. *Existence does not require your minds.* Since you have no solid answer to the reason for your existence, individually or collectively, you are left with just a Question. It is a single Question. Because the Question is all-encompassing, it does not necessarily need your limited minds to comprehend it, or even answer it.

People in countless disciplines pose an endless array of questions, trying to unravel the mystery of your existence, but there is only one Question. You dissect every angle, finding clues here and there, and unravel answers to your many diversionary questions, but are still left with no answer to the one True Question. How do you know what the Question is? The mere fact of your existence poses it to you. It is not a question that necessarily has words. But it involves two things – awareness and contradiction. Now stop. Think, for a moment about each one separately – awareness and contradiction – and what they mean in relation to your existence. You must reconcile them, if you can, and that is the purpose of this exercise.

The one True Question comes from the contradiction that is you. You are an animal species that has transcended nature. What sets you apart from nature is your ability to be self-conscious – to *know* that you are apart from it.

The sadness and travesty of your species is that the majority of you prefer to remain ignorant of the Question. But you are afraid of the answer. You know that the Question is there. You know how important it is. But you consciously choose to walk away and ignore it. You would rather seek power, prestige, entertainments, possessions, and economic production rather than the answer to the one True Question that matters more than any of those things.

Your True Self awaits you, but you fear it too much to confront it. Finding your True Self will provide the answer to your existence, but you prefer to sleepwalk through life, blind to its actual key. So when the majority of you die, you will have never lived at all. You would rather think about what could be, and analyze it to death, rather than actually *be* it, and experience it. When you finally pass on, you will have merely lived within the shadow of something greater that you will never know, unless you read this book and do the exercises.

The exercise for this lesson is simply to be aware of The Question. That's all. I have purposely made it simple because once you focus on The Question – and I mean truly focus on it – you will have your hands full with eliminating, or at least minimizing, all of the diversions that would otherwise keep you ignorant for the rest of your lives.

It is time to make a conscious shift in your life, using that God-like awareness that you possess. *Reconcile your contradiction.* You are apart from nature for a reason, and that reason has nothing to do with destroying it. For the time being, use your awareness to focus on The Question.

Your insights and realizations will help you in later lessons to make an even larger leap. The Lesson is over and the exercise should increase your progress.

PART TWO

BASIC LESSONS

Lesson Three

Focus

With the first two lessons over, you have laid the proper foundation for lessons of deeper magnitude. It is time to get into the lessons of power; those hidden or misunderstood for centuries. This lesson will sharpen your focus and improve the control you should have over your minds.

From an outsider's point of view, the easiest fault one can see with humanity is its lack of focus. There is almost no unity in mankind's actions. Little is being done on levels that would organize you in useful ways because there is no effective focus or unity. The focus you need is missing. You do not yet know how to use your minds effectively. You would argue to the contrary, puffed up with your bloated egos until you are blue in the face, but you are very, very far from using your minds effectively. The state of the world reflects this fact quite clearly.

If you knew how to focus your minds to the point that positive results could manifest around you, then the world would not be experiencing so many crisis situations and would be in a far better state of affairs. What I refer to is a collective focus, like the narrow beam of light from a high-powered source – one that will result from many advanced individuals acting as this one source (based on a common goal). They must first work closely and develop their united level of focus first. When a majority of you have developed your minds and focused them toward positive social change, then things will change for the better. A "common good" will result. However, your general focus is on personal gain, not on common good.

Positive, *effective* change will never be accomplished by leaders in politics or world affairs who work "for" you, because they are really not working for you at all. That is an illusion. It is a façade. You may get things from their efforts that serve you, and they will not hesitate to take credit for doing this "for" you, and make sure that you know about it, but this is merely a residual effect of a deeper agenda. You being "served" usually results in other interests being served in far better ways that you know little or nothing about. They are really working for themselves and the "special interests" that have gained their influence. The positive change you seek will only appear when you, collectively, do something about the problems that you are concerned with. You must stop complaining about the ineptness of your leaders because they cannot lead and most of them know that they cannot lead. Things have become too complex and too cumbersome for "leaders" to do anything effectively. It is up to you to develop focus and to do it yourselves. Your leaders will never admit this to you because they need their jobs and you are trained and conditioned to depend on them. Cast aside these "security blankets" and stand up on your own two feet for a change. Stop creating governmental puppet shows and become the main characters in your lives. Living your life is what you are here for – not to depend on others who pretend to be the main change-makers in your lives. Only you can make the best changes for you.

If you heighten your focus, true change can occur. The focus that you need to support the planet, rather than destroy it, is not being taught in your schools to any significant level, although this is beginning to change. Anyone who is a teacher and reading this needs to take note, followed by action.

Instead of nurturing the planet you are instead trained to become marketers and consumers of what is marketed. Cash flow is more important than compassion. Short-term profits rule the day over thoughtful, long-term agendas. Economic agendas crush ecology, and will thereby crush you out of existence. The writing is on the wall, but you prefer instead to direct your catchy little jingles into the minds of the masses, turning them into automatons that sleepwalk into your Wall Marts and walk out with empty wallets. Everyone is happy – or at least believes they are – as you collectively sink deeper into your economic quicksand each and every day, while lessening your ability to escape the inescapable. It is like the Venus flytrap. The fly is attracted to the sweet nectar, but once it closes on the fly, there is absolutely no escape.

The nectar of material possessions has you mesmerized. Your attention spans are very short, you do not see beyond your immediate needs and short-term desires. You are trained to respond immediately to clever marketing; to be consumers. You are a society of Pavlov's dogs and the ringing bell is the television and its commercials, or the Internet and its pop-up ads. Although you are higher primates, you have mastered the art of making your fellow humans respond to such mindless and insulting forms of stimulus, that you are, in effect, nothing more than salivating, trained animals, functioning below your true potential – Pavlov's dogs, with very little control over what you consume. Reason will tell you to consume far different items and to use resources that make more rational sense – but you are not rational creatures. You are first and foremost emotional creatures, trained to act on your immediate desires rather than rational thought. You currently have no clue as to how to develop meaningful thought patterns – patterns that will help you achieve the focus needed to become a functioning part of nature once again, rather than its worst enemy.

Thought is the father of action, but you have surrendered your thoughts to others – who thereby *control* your actions. Your actions are important. They are reflected in the physical world after they are developed mentally. Control of your thoughts lead to greater control over your actions and therefore greater control of the things you manifest in the physical world. But right now you have very little control over your world because you have very little control over your thoughts. They are scattered all over the place. Few of you can hold a powerful thought long enough to cultivate it properly. For example, you may go to a job interview but forget what to say when the crucial moment comes, despite planning for it so carefully. You concentrate on what must be said beforehand and only manage to express a semblance of it when you need to. And when the moment passes, it all disappears from your mind. Most of you fail to improve your minds and perfect your delivery, but instead search for more encounters. You *refuse* to focus. Your confused inner whirlwind – that jumble of thoughts that compose your mind – moves to another thought and leads you astray from what just occurred. Far more important things than jobs, including life and death issues, suffer the same fate in your world because this is a *pattern*. You are trained to repeat it.

You are trained to remain unfocused and to be consumers. Studies show that most things purchased are bought out of impulse. It is called impulse buying. Marketers want you to have a short attention span. If you have a short attention span 1) you will be more inclined to buy impulsively and 2) you will be less likely to examine your purchase later and realize that you did not really want or need the product. Rather than return the product, you will most likely move on to other things. Reflective thinking is not encouraged in this materialistic world of consumerism. It is frowned upon because it makes no money. Reflective thinking will provide you with long-term solutions rather than short-term profits that destroy your environment. Reflective thinking will present things to your people in a more thoughtful, caring way. Remember the last time you did something really stupid? You failed to reflect properly before acting. You have all had such moments. Chances are, you said, "I just wasn't thinking!" Your entire society acts in this way – it is *trained* not to think.

Right now your entire society is geared toward gizmos, gadgets and gimmicks. Just look at the primary means you use to sell your items throughout the world. The television commercial. They reach more people to sell things than with any other method. At the same time, they broadcast your ignorance and stupidity to more people than with any other method, presenting gizmos, gadgets and gimmicks that most of you do not even need. Your commercials show people acting stupid and ignorant. That is how low you will go to sell cheap items of little usefulness – items that may get used only a few times before they break from poor workmanship or begin to gather dust after the immediate novelty has worn off. Millions of desperate people need your help, but instead of creating value between yourselves through compassion, you must peddle your junk, giving it the value you have lost among *yourselves*, by using clever jingles. You degrade yourselves in the most blatant ways in exchange for the almighty dollar. You must ask yourselves this question. Is it worth it? Right now, you think it is. But it's not.

Your advertising tells others what they desperately need, but most of it is unnecessary junk. You are training each other how to think, and what to think. You are creating desires that aspire to materialism rather than higher ideals. More importantly, your television advertising is the reflection of your consciousness. Instead of looking into a clear pool and seeing a nice reflection, you are staring into a cesspool. It is a cesspool of greed, waste and ignorance. *And you can't wait to jump in.*

If people are trained in higher thinking and in developing moral character, they will purchase what is right for them without being tricked with gimmicks. They will see clearly that a cheap gadget is nothing more than that, and will instead get useful items. With useful items as objects of focus, your actions will change. Thoughtful people using practical items create true progress rather than destructive habits that destroy the environment and erode your moral character. You will experience less theft, murder and crime because the objects of your desire will not be primarily material. With the proper mindset, there will one day be a useful balance of your material needs coupled with caring, compassionate conduct.

This can never be accomplished, however, without better focus, clarity of mind and longer attention spans. Today you are intentionally barraged with short, snappy attention-getting jingles and lightning-quick, spontaneous images. The goal is to have you make an impulsive buying decision *without thinking*. You are being trained not to think, but to *react*. Like Pavlov's dog, when they ring the bell, you must salivate and buy the product. Immediately. Do you want to live your life this way? Or do you want to think for yourself – to *really think for yourself* – rather than being told what to think, and buy, and do for the benefit of others? You have allowed your puppet-masters to turn the switch "off" on your very souls, and have voluntary chosen to become mindless automatons for *their* benefit – not yours.

Focus. That is what is missing in your lives in a critical, primary way. I will teach you to focus in the following short and powerful lesson.

You are always looking "out there" for every single answer. Your focus is always "out there," at objects, because you have been trained from birth to focus on the outer world alone. You must assess the things in the outer world as being a threat or not and, if not a threat, to decide whether you want to possess them or not. It's a very simple process.

Focusing in an inward way is foreign to most of you. However, it is essential for your spiritual growth. More and more people are discovering the importance of spiritual growth after realizing that the materialistic, physical world does not enrich them enough. Something is missing, and they begin to search for it. It is a spiritual yearning of some kind, and it begins to stir only when you are ready. It is not to be found "out there." Their main question is, If it's not out there, how do I get to it?

Let us begin. You must realize, first and foremost, that you do not "get to it." Your inward focus is not something to chase. Nor is it something to learn. You already have a powerful inward focus. It is already there.

Exercise

Finding Your Power Point

This is a simple mind/body exercise. It starts with your state of mind. Most people are not satisfied with the present moment and are always planning for the future or dwelling on the past. They are usually trying to figure out what they need to do in order to improve their lives. They never live in the moment, but instead are always planning to make their future moments better.

Forget about the future. Your true power is in the present moment. That is all you have – just this moment. Stop and realize this fact. *You have nothing else except this moment.* Once you realize that, your power is within reach. You get so caught up in plotting and planning in an effort to make things better, that little or nothing gets done! The energy is lost in various forms of planning rather than one good plan being executed properly, with focus. After falling short you then go back in your mind to the past, trying to figure out what you could do better next time. The moment is always lost, because the art of being fully conscious in any given moment has also been lost.

Most people are too afraid to act when the crucial time comes, so substitute planning and remembering in its place. When "Plan A" fails (because you were too busy thinking rather than acting), you move on to "Plan B," which fizzles, followed by "Plan C," and the pattern continues from there.

Society demands mediocrity and for one to break out of this mold requires great talent and discipline. Most people have resorted to fantasizing about success rather then actually achieving it. You resign yourselves to failure in the actual world, and create a substitute "fantasy world" that soothes you as a substitute. Television brings the fantasy right into your living rooms, bringing life to your fantasies, and keeping you pacified. Your true power, as a result, gets buried and your true potential dies.

You should never allow your soul to suffer in this way. You came here to *live*, not to live through others vicariously, or through self-imposed fantasies created to provide you with at least some sort of outlet.

A strong *focus* is your answer; it is your way out. When you dwell on past memories or future planning, there is a certain center of power within you that is empty. That power is only with you when you are focused *on the moment*. Being fully present fills you with spiritual power.

You can either memorize the following paragraph or have someone read it to you for guidance.

Close the eyes. Focus on the moment, which is all you really have, but place that focus in your solar plexus. Be relaxed, breathe through the nose and out through the mouth. This will charge your point of power with energy and you should feel it as you progress. Notice how calm and aware you now are. A certain strength and confidence should also be present. Now move your awareness up into the center of your chest and repeat the process. You should feel even more power from this chakra, or energy center. You have numerous chakras in your body, but these two specific areas are where your main power centers reside.

Now leave "the moment." Go back into your mind; into your normal mode of planning and/or remembering. Note how this feeling of power, confidence and strength leaves you. That is how most people function – *without their true strength*. Now that you know how to access it, begin to use it in daily life. You have to stop and make a conscious shift to activate your power, but you will find that it is well worth it. It is easy to forget, and to stay trapped in the mind, but accessing your power in crucial situations can provide you with entirely different outcomes. It can change your life drastically, if used wisely. When you tap into the moment and access your power, the mind also becomes very calm. Did you notice how calm the mind becomes? When comparing this power point to normal consciousness, it becomes clear how much the mind jumps all around when you are not grounded in the moment – operating almost in a panic compared to when you have your full power intact.

You must stay in touch with your power often so it becomes second nature. It will take time to do this. When you access it, it can be increased with the change in your breathing, as described above. Accessing your power through inner focus may create a distraction for you from the outside world. The trick is to remain conscious of the outside world, but view it with your inner mind – the soul – rather than the normal, limited form of consciousness.

If you fail to experience anything but a subtle shift from this exercise, stay with it. Keep practicing it because you are not accustomed to this shift and with accessing your power. But when you keep going back to the well and dipping your "bucket" of awareness in a little deeper, you will come up less empty each time. You will eventually access the very waters of life that will strengthen and nourish you in ways you never dreamed of.

When you depend on the mind alone, you are empty. Without your true power the mind is a snare, it is a trap that can suck the power right out of you. Most people believe the mind to be their greatest asset, while those with wisdom can see its faults more clearly. The exercise allows you to oversee your own mind from a higher vantage point and prevent some of the mistakes that you would otherwise make spontaneously. It is a very easy exercise to perform. Practice it a few times, with one good session now.

If you have successfully done this exercise, your ability to focus has now been increased. You can oversee your own mind from another perspective and thereby exercise more control over it. The next time you encounter any event, meaningful or not, access your power and be in the moment. Keep practicing it; you will get better each time and your results will improve – especially with events of importance. Using this focus to change your life and the lives of others is now up to you.

Lesson Four

Wisdom

The intelligence you possess has brought you as far as it can go before starting to become detrimental. You must begin to use it differently. It may now be brought forth to become wisdom, *if* you are ready to use that wisdom properly. I have determined that you are ready to hear my words and cultivate your wisdom. That is the purpose of this lesson.

I can help you to cultivate your wisdom but what you do with it, however, is the important part. But first, we will start with basics. You must begin to grasp true wisdom and understand its source; then you must learn how to use it properly. Your knowledge has been a weak substitute for wisdom, when wisdom has been needed. Knowledge and wisdom are two different things. All the knowledge in the world, no matter how powerful, is useless and even detrimental if it is not used with wisdom.

Throughout your history you have used wisdom well in a few cultures, but it never lasted very long. Most of the time, throughout your history – and especially today – you have mistaken your advanced intelligence for wisdom. They are as different as an insect on the ground and a giraffe that can view things from above the trees. An insect can cover a lot of ground and learn the landscape perfectly, but it has a limited view of the world. A giraffe can see for miles. A predator has far more difficulty attacking a giraffe. Insects get eaten all the time because they don't have any idea that they're in danger and cannot see their enemies coming.

Humanity is in the same boat. Your intelligence works well within your limited view, but the world truly does operate at a more sophisticated level than you can comprehend – believe it or not. You think that you are safe in the way you conduct yourselves, but you are in far more danger than you know. With each passing day this danger increases.

Therefore, Lesson Four involves the cultivation of wisdom. There is no point in you learning anything from this point forward unless you can exercise wisdom in using it. Some modern esoteric schools recognize the value of wisdom teachings as their ancestors did, and teach them only at higher levels. The vast majority of schools disregard wisdom teachings altogether, maintaining their focus on knowledge alone. This has been one of humanity's gravest errors, so we are about to correct it here. All of the great technology seen in the world today is the result of your expanding knowledge. Yet, with all of your amazing inventions and gadgets you are spiritually bankrupt. You still grope in the dark, not knowing who you really are, how you got there in your world, or where you are going. Your various religions provide solace to some degree, but record numbers of you have turned away from your religious institutions, intuitively knowing that some other answer must exist that is not so outdated. Your religions offer knowledge, some of it spiritual in nature, but they do not train you in wisdom. Only you can do that. Faith is not wisdom, and it can be blind. You are beginning to look for answers that move beyond mere faith and make more direct, rational sense to you. There is nothing in your world, *nothing*, that trains humanity in the cultivation of wisdom that is accepted by the masses. The masses are intentionally kept from developing wisdom – to your own detriment. You spread knowledge around, thinking each time that more of it will finally provide you with the right answers. It does give some answers, making you believe that you are on the right track, but knowledge feeds the ego and the ego is never satisfied.

Wisdom does not feed the ego. It gently sets it aside, bringing in a higher power that knows better than to add fuel to an insatiable fire. A deeper part of you can be fully satisfied. But you are trained to believe that this deeper part of you is the ego, and the one thing that truly matters, so you continue to try to satisfy it. You have identified yourself with the ego, which constantly cries like an overgrown baby for you to fill its every need. It always wants something, and you are always more than willing to cater to it because you believe that doing so is the only game in town. It is *not* the only game in town.

What fuels your constant desire for things is the ego. And this ego is connected into your physical brain through the neuro-pathways that you have created since childhood. You have been trained to operate in electromagnetic ways that involve your ego and the circuitry in your brain. The ego "owns" your brain and does not want to surrender it.

Most people hear a constant chattering of thoughts that express themselves in their minds. Listen closely to what it tells you. It is entirely based on its desires. They are not "your" desires. You are just being told that they are yours. And you believe it. It has made you its slave, based on its desires. You are nothing more than a peon to the ego; its pawn. This is news to many of you. You have always thought that *you* are in control of your own destiny. Think again! You have a stowaway on board. It has jumped into the helm and is suddenly piloting your ship. Because it says it is in charge, you have accepted it without deeper examination and have surrendered your power to it. Pay close attention to your thoughts, and as each one arises ask yourself who is doing the thinking and expressing. Is it really you? Or is it some lower part of you that has somehow taken over? It has wedged its ugly foot into the door of your mind, and you have never been able to get rid of it. It has convinced you that it is really you, but *it is not you*! It wants you to believe it, because if you fail to do so the ego believes that it will die. It will not die, but it is like a child and thinks that it will. As a result, it clings to your mind tenaciously like a spider with its prey and refuses to let go. It refuses to let the higher part of you in, which carries wisdom, because it is addicted to knowledge. Knowledge, like the ego, rules you and therefore rules the world. That is why there is so much suffering, imbalance, and lack of compassion. The ego cares only for itself, and allows all others to suffer at its expense. Wisdom operates with long-term foresight, compassion and mutual respect; while knowledge feeds the need for short-term gratification and requires constant defensive adjustments to protect itself. Wisdom needs no "protection." It does not need to defend anything.

The cultivation of wisdom was once part of the ancient mystery schools. They have been long abandoned, due to barbaric takeovers or cultural shifts into more materialistic pursuits. Today, you are involved in so much self-importance that you can no longer be bothered with restoring wisdom. Much of the teachings have been lost or intentionally burned in order to protect religious dogma. This dogma is still holding on today, but its influence is slipping away. The world has failed to offer access to true wisdom, or at least the methods to reach for it, so let us explore how to do that together. A door has opened and the time has come to enter it.

The first step toward wisdom is silencing the chatter of your mind. In India, where wisdom schools have been kept alive to a small degree, they call it the chattering of the monkey, or monkey-mind. A monkey has little foresight and is quite vocal – it just reacts to its surroundings based on immediate need. Since you share over 99% of your DNA with chimpanzees, this "monkey mind" comes as no surprise. However, part of you is progressing in a spiritual way, and that is the "opening door" which I have mentioned. It can bring you beyond the monkey mind. Collectively, you must walk through this door and leave the ego behind. If you (or your minds) just stand and scream at the new passageway that is opening up to you without actually *doing* anything, then you will remain in your lower state and the world will continue to deteriorate all around you.

In an individual sense your mind operates like the worst nagging housewife or nit-picking husband the world has ever known, but you have gotten used to it. It is "normal" for you. For most of you it is the only thing you have known, with all other options blocked off and foreign to you, so it is accepted. However, there are those who have silenced this constant dialogue and have achieved clear access to wisdom and, on better days, bliss as a common result. To a tranquil, quiet mind, your "normal" constant chatter is found in a mind that functions on the brink of insanity. And as one thinks, one acts. Therefore, the world is operating at the brink of insanity. Certain parts of society have sunk so far that insane ideas are accepted as the norm and adopted into your daily lives. Sometimes, just sometimes, a clear thinking person will step forward and urge a corrective action, which is sometimes made. But the vast majority of your actions goes unchecked and continues to damage you and your environment.

This is where the previous lesson, focus, comes in. Are you in charge of your life, or is it this monkey mind that is ruling you through its desires? You must take control of the ego-mind by getting completely focused on specific things that *you* know are deeply important. Your minds are not focused, but they can be if you take charge of them. Your mind will start to listen to you and work for you, instead of against you, if you take it and tie it to a post (symbolic of one main, positive focus) so it cannot escape. Once it realizes that *you* are in charge, it will begin to quiet down. The key is to use your focus with wisdom and persistence. Very, very little of this is being done today. You run in circles, concerned with the most trivial garbage your minds can conjure up, while convincing yourselves that these things are "important."

Then you set out to justify this foolishness. Your minds are dualistic in nature and therefore you are always looking for a "bad guy" who will justify your foolish actions. You are always vulnerable to losing whatever credibility you have unless you can, sadly enough, cast blame onto someone else, whether they deserve it or not. Almost all of you allow the ego to run your lives. You are therefore a paranoid race that is in constant denial. You claim that you are in control only to the extent that you can maintain the *illusion* of control. The reality is that almost *none of you* are in control. It is all an illusion. You accept the illusion because in exchange for your trust in the ego, you receive a false sense of security. The bottom line is that this form of thinking is not "normal" at all. It is warped and rather primitive.

When dualistic thinking is overcome, however, all illusions vanish. If you were to accomplish this, your left and right sides of your brain would work together, rather than in polarity. You would begin thinking holistically and therefore start acting holistically. Your "bad guys" would begin to vanish and you would dramatically increase your good will and aid to others. You would start caring about each other and the planet more. This shift is beginning to happen – and *wisdom* is the ultimate result.

Your biggest illusion, even beyond the "control" illusion, is the "conquering" illusion. This one has done the most damage and holds the greatest need for being overcome. Your armies have been trained for centuries to overcome one another and you are compelled to conquer everything in sight on a personal level, no matter what it might be. It really doesn't matter, as long as you can gratify your egos. This is the sickness of mankind. You believe the only way to be safe is to control others, and the only way to control others is to conquer them.

Your primitive instinct for self-preservation has created this ego-based compulsion. When you lived in caves self-preservation was needed. You now need it less. Because of this, the ego fears losing its sense of self-preservation, so it has turned into a compulsion. It is a sickness. If you can control more and more of your world and each other, the ego believes it can preserve itself. The opposite is happening because a preserved ego cannot function in a world that has thrown the joy of simplicity out the window.

It is time to move beyond your primitive instincts. Right now, it is like you are trapped in a room (your world) filled with all the junk that you have accumulated that has only been made "necessary" by the machinations of your minds. There is a deeper part of you that can dispense with all the frivolity because it can finally be *recognized* as frivolity.

Examine how often you engage yourself in the pursuit of entertainments as opposed to helping others and changing their lives in a positive way. So much can be done to reach out to others who really need your help – but instead you are watching forms of entertainment that have no bearing on you personally, and have little ability to improve the lives of anyone, including yourself. But it is addicting to the ego. You become titillated for a brief moment and because that isn't enough, you continue searching for other frivolous forms of entertainment. In the meantime, your entire social structure and the environment is deteriorating all around you, and you wonder why. The technology that society pushes on you literally *demands* that you engage yourselves in the pursuit of entertainments. Those creative individuals who may have something to offer the world on a large and meaningful scale end up becoming entertainers rather than humanitarians. It takes a special person to become a humanitarian, largely because entertainers and mainstream careers offer more money.

When you get your priorities sorted out and begin rewarding humanitarian action, then things will begin to shift. A new paradigm is on the horizon and is truly coming because you are recognizing the proper directions you need to go.

You need wisdom to bridge the gap. Your high degree of knowledge has its positive aspects but keeps you trapped within the ego, which is selfish. Therefore, humanitarian aims will remain limited without the cultivation of wisdom throughout your culture. When this wisdom comes, it will serve as an awakening on many levels – not just with humanitarian work.

For example, your mind operates using dualistic thought patterns, despite the fact that it is infinite. Your mind is your main tool, so it is limiting in the way you use it. This is like using only a hammer when an entire toolbox with every kind of tool is hidden in the recesses of your minds, complete with building materials and floor plans, waiting to be opened up in your mind to access enormous possibilities. Instead of using a limited tool and ultimately destroying your surroundings with a hammer, you can build an entire universe. With wisdom.

You have begun to access your wisdom to the extent that you are being forced to. Wisdom cannot be hurried, or approached through desperation. It must be cultivated. Very few of you are attempting to cultivate wisdom. The wisdom you now have will never be enough to help you, to *truly* help you, unless you can collectively adopt new habits. You cannot wait for others to do it for you. It is time for each of you to step up and take responsibility for your own lives, rather than turning over that control and trust to others, through the dualistic illusion just mentioned. You must do away with the illusion and not only think, but *act* holistically in every area of your life.

Humanity is afraid of the fall of civilization as you know it, and you are aware that it is immanent. Yet you cling to it, desperately, hoping that the problems will go away by themselves. They will not. With wisdom, you can install the safety nets, but your greed and runaway desires fuel the dying paradigm just enough to convince you to put it off and do nothing.

You have outsmarted yourselves. You have painted yourselves into a corner and the corner gets smaller every day. You still have room to maneuver, but if you wait too long, your maneuvers will not be effective enough to free you.

You are not going to develop the wisdom you need any time soon, but you are making progress in this area. I am referring to a major transformation. For you to salvage your rather dire situation it is essential for those most visible and influential among you to show others the way. This will require boldness and bravery, because when you begin to operate in a realm of more awareness it may affect the economic power structure – a power structure that is damaging your world in countless ways and needs to be altered. Those most visible among you can best set the example and begin the change. And if you who are reading this is not "visible" or influential right now, it doesn't mean that you will never be. Find those who are instrumental in this new awareness and align yourselves with them. Officially. Make yourself available and at their disposal should they "need anything." Make sure that you are not only known as being available in this way, but that you actually *do something* to forward whatever important cause you have chosen. If you do this, your life will take on an incredible new meaning – and all that is needed is for you to sacrifice a few entertainments.

Those with true wisdom must be sought out and consulted. They must not have, under any circumstances, vested interests with any organization rooted in the operational structure of this dying paradigm. You must consult with those who are detached enough to make clear-headed choices that will bring down the beast as gently as can be expected, by listening to them and acting on their advice. The economic power structure, however, has so far refused to slacken their reins enough to allow any change whatsoever, unless it is forced upon them. They have pulled their strings and manipulated world finance masterfully, over the years, with their advanced intelligence, but the lack of wisdom is their potential downfall. Their own personal comfort is more important, maintained with great intelligence, manipulated at the highest levels, while more and more of the world falls into discomfort at their expense. They plan to transition into new operational territory using the same mindset and without surrendering any power, but the power must fall into other hands once and for all if you can ever hope to progress.

You have allowed them to insulate themselves from you, and their money and power protects them. They wish to provide positive change on their terms, since they are in control. They are intelligent enough to continue their power, but not wise enough to use it properly, as evidenced by the state of world affairs.

It is up to you to recognize what needs to be done, and act with wisdom – for only with great wisdom will truly positive change occur. Using people who have this wisdom is essential. Those who are considered to be spiritual masters are, in today's world, not counseled too closely by world leaders. World leaders are more interested in economic gurus, not spiritual ones. But spiritual gurus understand things on a far deeper level, a more meaningful level. They are not focused on short-term economic gains, but on the long-term spiritual realities of the world and how you should be operating within it in a holistic sense.

Yet, the world moves too fast for such contemplative thoughts. It wants results *now*, and doesn't have time to plan or cultivate deeper truths that take time to develop. I am here to tell you that there are no shortcuts. Inner work, through reflection and meditation, combined with life experience and a reasonable degree of intelligence, brings wisdom. You have been using faster means with instant gratification in mind to operate your world. And soon you will have no choice but to turn to spiritual men in robes to save you, rather than men in suits and ties.

You will suddenly realize what a fine wine can taste like because you have never aged it properly. Learn to meditate. Look deeply into your mind and take steps to make it still. It does not happen overnight, but with years of meditation and developing your inner powers and your insights, amazing things will unfold for yourself and those around you. You have been living in darkness, and I am showing you how to crack open the door just enough for you to realize that light is on the other side – and that advancing in this direction is the proper course to take.

Exercise

Accessing Your Wisdom

This exercise is a continuation of the previous one, called Finding Your Power Point, so refer back to it on page 44 if needed. People believe that wisdom is something gained through experience. That is not accurate. You already hold a great deal of wisdom within your true self, or your higher Self. The previous exercise has given you access to your power point. Within that power point resides connections to your higher Self and from them you can tap into your wisdom. The previous exercise was the easy part. You practiced it a few times. The hard part is integrating it into your daily life so that your wisdom effortlessly shines through and you become more and more of who you really are.

Your higher Self looks down upon you from afar, although it is still connected to you and remains an important part of you. Knowledge is gained primarily through experience; wisdom you already have. You must learn how to reach in, pull it out and *use it*.

Your exercise is to enter into your power point (repeat the last exercise) before you engage in any important activity and, as a continuation, *use the wisdom you have to guide you*. Stay focused on your higher intentions and the guidance you are connected to. You will act with more rational intent and calm reserve. You will not rush into the situation without thinking clearly about your actions and their consequences. You will have better insights into your decisions and your next moves will play out in ways that would have not otherwise appeared.

It is very, very easy to slip out of your power point by getting caught up in your normal, imprinted thinking patterns. As a result, you will lose your focus. Using this focus employs your wisdom.

This is, in effect, operating from an enlightened or semi-enlightened state. It is not easy. But remember, as far as wisdom is concerned, you *already have it*. It is not gained through experience at all – but is most definitely used through your experience once it is "grasped." Most people have this wisdom, but rarely use it in their lives. With this exercise you can employ it daily, if need be. It is like having a car that you keep in the garage and never drive. Get out and *live life*; drive it and *experience it* every day because you now know how to use the vehicle. With wisdom. You will go places that you will have never dreamed of. As a result, you will sometimes take people with you that will be forever grateful for taking the journey, or for you showing them the path. That is what this Lesson should accomplish.

Lesson Five

Compassion

This lesson will connect you emotionally to the natural world and the people who share it with you. You – meaning humanity – have disconnected yourselves from a proper emotional attachment to your natural surroundings for the sake of material and financial gain. A certain coldness and detachment is required by the ego to place yourselves into positions that make money through exploitation by various means. This has led to your detachment from the very life-cycle of the planet itself. You must be brought back, emotionally, if you are to have any hope of surviving for a reasonable duration in the future. That is the purpose of this exercise.

Millions of people are starving at this very moment and little is done to help them. Many are being killed in wars or terrorist activities around the world. This violence is not only government sponsored, it takes place in your inner cities among gangs or surfaces between families and between individuals in violent, domestic disputes. Humankind fosters and breeds among itself a war-like mentality. On the surface this does not seem to affect you, so you take an "out of sight, out of mind" approach to it. You sweep it under the rug unless you are confronted with it directly. You don't understand that this violent mindset has created your entire planetary environment and it affects everything. Your environment, therefore, has a very low, negative vibration that every other intelligent life form in the universe wants nothing to do with. You are not alone. You are being *left alone* – intentionally, and for very good reason. Collectively speaking, you are not ready to interact with intelligent life from elsewhere because, despite what you think of yourselves, you do not act with intelligence at all. Whatever intelligence you do have is being largely misused. It is blinded by ego and fueled by greed.

Much of the world lives in filth and poverty while others not affected ignore the victims like they do not exist, living in extravagance and wealth, gambling their money away or misusing it, while the less fortunate starve and suffer almost before their very eyes. What is it going to take to *open your eyes...* so that you will act with compassion?

One common trait among dangerous psychopaths is that they have no compassion. The majority of mankind acts with no compassion, so therefore, mankind, as a whole, is psychopathic. The state of the world proves it, but you refuse to see it or make efforts to change it.

Go back, read the previous three sentences one more time, stop reading immediately after this for a few moments, and think about it. Pretty scary, isn't it?

By contemplating this properly, you have begun to awaken. You see the world for what it is, and must now design your life to correct it – at least within the areas that you can control. Compassion is your key. It has always been your key. Your greatest people were also your kindest and most compassionate. Yet you teach in your history, political science, social studies and most classes around the world the basic idea that "might makes right." Your focus on history is one of conquering and suppression – yet there is an equal history of humanitarian work that gets swept under your intellectual rugs. Therefore, you train your people from a very young age to be – to put it bluntly – violent, crafty and opportunistic hooligans. Those who devote themselves to lofty works are often the victims of parasitic wolves who wait in the shadows for their chance to steal the prize, like the hyenas in Africa, who gang up and steal a meal worked so hard for by a lion or leopard that was needed so desperately for the feeding of its' young.

There are far more wolves waiting in your shadows because you have taught them to hide and wait. To hide and steal – rather than develop their compassion so they work in service of others, make this world a better place, contribute something positive, develop their self-esteem, and feel, or rather *know*, that they have accomplished something *useful*.

It is far easier to revert to primitive instincts and act as predators than to live with kindness, so mankind most always chooses the easier and more brutal route. Callous actions produce faster results, fruits for the victors, and innocent victims. However, if you develop wisdom, compassion will follow and more of you will be happy. When operating from wisdom, more of you will work together and benefit together in synergistic relationships, rather than pitting yourselves against each other with the false illusion that you must create a loser in order for you to "win."

Those lacking great wisdom may still cultivate compassion, because compassion is an *emotional* act. Sensitive people with reasonable intelligence have displayed amazing acts of compassion – and this in turn helps to *develop* their wisdom. Something springs up inside of them as a result. If you give up a material possession or funds, or your time and hard work to help those stricken in a disaster, with little thought of material gain, your heightened sensitivity does not perceive this as a "loss," because you have created a position in the world for yourself that moves things forward and makes it a better place. In direct response to witnessing the fruits of such actions, the soul advances and wisdom results.

When new opportunities arise that may have similar results, you recognize the opportunity more quickly, as do others, and results on a collective scale do not need to germinate as long. This process is part of the awakening process that you are all experiencing, and I am trying to guide you through it. Compassion is a major component in your development; it is a measurement, a barometer, of your progress.

Do all you can to create compassionate action with yourself and by others. It is the food for your soul. The growth and strengthening of the soul is a major reason why you are here. Most of you are completely ignorant of this fact because your attention is directed toward "things" rather than actions. You do this out of habit. Your physical surroundings have become more important than your own personal and spiritual growth. You jump like a fox trying to reach grapes that hang just out of your reach, thinking that material gain for yourselves will make everything right. Many foxes will jump and get a few grapes at a time, never satisfied, until they die of exhaustion. Some can be clever, however. If they should climb the tree and stand on the branch containing the grapes, it will drop low enough to the ground where the other foxes can grab them. In gratitude for this, it would only be natural for those in receipt of this gift to save some for the fox who gave of himself for the others, with not necessarily expecting anything in return. You need to stop thinking in greed-based modalities and graduate to the next level. You are showing signs of this happening, but your consciousness must jump to another level and spring you into another paradigm. Your current paradigm is no longer serving you. Until you abandon it, you will continue to experience degradation in all walks of life. The positive, uplifting events that you will experience will occur in direct relation to holistic thought and action.

Exercises

This exercise has two parts, the first is called Transcending Duality, with the second being Great and Small. If you do them both they will add greatly to your life.

It is easier to act without compassion and harm others if you have not identified with them on an emotional level. As soon as you sit down with your enemy and get to know him, he is identified with you and often ceases to be an enemy.

You are all connected. In fact, everything is connected in more ways than you know. It is within your minds that enemies are created because you suffer from the disease of dualism. You need conflict; you need "bad guys" to fight against so you can build up your egos and your self-esteem, giving you the illusion that you have accomplished something great or have overcome diversity. Your enemies constantly change, based on your self-centered, immediate needs.

Some enemies are in fact malicious. When you defeat those few enemies that are truly malicious, it does in fact become a character building exercise. But the human mind gets so addicted to "winning" that it will create conflicting scenarios that do more harm than good. You like to build up your enemies to appear as bad as possible, so if you defeat them it will be a sweeter victory to the ego and if you lose, you can blame it on the great "evil" you were up against.

Transcending Duality

Identify someone that could be considered an enemy of yours. Exclude anyone who might be mentally ill; it should be someone who has all of their faculties. Exclude those who have done irreparable harm to you, as these situations could well be beyond repair, depending on the level of damage done. Good choices include any competitive rival you personally have or have had in the past, members of sports teams that you have always disliked the most, relatives who you have never gotten along with, or those who occupy a job that you hate and would never do, including those that involve questionable moral character.

Your job is to *not* make friends with this person. It will be hard enough forcing yourself to speak to them or to research them. Keep your distance to a comfortable measure, but try to learn as much as you can about them. Do some research through public or private resources. Try to find out what motivates them and why they do what they do. You could also include an interview (recommended), if safe to do so.

Your task, upon finishing your research, is to write a one-page report about their goals, attitudes and motivations. Request a few minutes of their time on the phone, in person or even through email. Let them know that you are writing a human-interest story about them and you chose them because of their accomplishments in life. Tell them you are writing something about them as an assignment (for a class if that is the case), because you believe you might learn something about yourself. This is an honest approach that would do well without revealing your aversion toward them. Do not tell them they are undesirable unless it is clear already, from past encounters, and it exists only in a minor way. Do not choose someone who could be excessively rude, hostile or violent toward you under any circumstances, based on past experience. You want helpful information – not to "stir the pot" or create trouble. If you end up with more than one page written, that is fine. The point is to gather as much information as possible and to study it closely. After you study what you have discovered, there is one final task. Go and observe them, if possible. It doesn't matter if they are a celebrity, sports figure or a relative at a family get-together. Observe them closely and notice if you see anything different about them.

You are not asked to make friends with this person and it is not expected that you will. There always exists the possibility that it could happen, but it is not an ecouraged goal or an expected result. Pay attention, however, to how this person perceives you, in addition to your own observations of them.

Great and Small

Part Two involves your home or work place. Remember, everything is connected. All of life is sacred. When you find any form of life in your personal living or work areas, no matter what it might be, remember that it is sacred. God created all life and when you kill insects or other creatures you are disrupting a natural harmony among all living things. A smaller being trapped in your home did not intentionally choose to come inside just to annoy you, although that is how most people react. Its' harmony with nature is already disrupted because it is in your home, so it is just as unhappy about being there as you are to find it.

When an insect is found in your home, it is the first instinct of many to just kill it. That is the height of ignorance and is a truly callous act. The first thing you must think, right away, when you spot a tiny creature in your home or workspace, is that it needs your help. It needs to be out in nature again and to be happy and healthy. It is a *privilege* of yours to help this creature by returning it to nature. It is not a privilege to kill it.

Your life and this creature's life are both sacred. Life is life, whatever the form. Despite your inflated ego and superior intelligence, you do not have the right to kill something simply because their life is a "minor inconvenience" to yours. If far superior giants suddenly took up habitation on the Earth and you did not know, from moment to moment, when a giant newspaper or other unknown object would come crashing down out of nowhere to crush every bone in your body, instantly, then you would begin to experience the kind of paralyzing fear that you instill in these defenseless creatures who do not know how they got into your home any more than you know how you got here on this Earth.

In God's eyes, especially in relation to the balance of nature, you do not sit well in His graces. You are not as high and mighty as you think you are. Your job, therefore, is to extend compassion to smaller creatures and help them when you can. It is a *privilege* to do so. All of life is sacred and connected, so you must do your part to maintain it.

Nature has a certain vibrational flow that can be felt when you are in tune with it. It cannot be explained very well; it must simply be experienced. When it is experienced, a certain conduct – far different than what you currently exhibit – results. Mankind has lost this connection and therefore, its' proper, balanced conduct toward the natural world. This lesson will put you on the path toward restoration, but you must put it in practice often.

Lesson Six

Mutual Respect

This lesson should improve your respect toward others even if you think it could never be improved. As a result, your interrelationships will impact the world around you in an entirely different way.

The human mind has been amazingly bigoted. Throughout history you have looked for the slightest differences in your fellow man in order to launch brutal attacks on him. Whether it be color of skin, a difference in belief systems, being from a different country, speaking a different language, having a different sexual preference, or wearing a different style of clothing – it doesn't matter. As long as you have something to use to create an "enemy," you will. Somehow, in your narrow, egotistical points of view, you believe that everyone should be cast in *your* image, rather than God's. If someone is not exactly like you and fails to conform to your idea of reality, then they become an "enemy" and must be subjugated or die. You somehow form the opinion that others are not fit to live if they vary from your lifestyle or belief systems. And then you go out and enlist God (within your own minds) as supporting this warped mentality. You categorize these others as being sub-humans, who must be treated in cruel and inhuman ways so as to reflect this manufactured, lowly status. It is an amazing example of ignorance that gets repeated over and over again, in different forms, in different times, in different places.

The worst form of self-elevation involves the degradation of others, especially by those who are otherwise incapable of elevating themselves. As an alternative, they must psychologically degrade others because it is the only other available means left to increase their value and self-esteem. Your small minds use this tactic repeatedly, but to those watching from on high, it earmarks you as ignorant fools who are not yet mature enough to create a lasting and meaningful culture. That is why your cultures fall or get destroyed. For us, who monitor and assess you, it is like watching a bad movie and we can only stand so much.

Cultures with forced divisions among its people breed animosity, not harmony. You are only now beginning to understand this. Only recently have you allowed those of darker skin to live as free people rather than as forced slaves, and your females to hold an equal place in your societies to the males. This, however, has only happened in some of your countries. In others, the problem still remains and you have far to go.

You must learn and exercise mutual respect among all people. Only animals gang up and exploit those perceived to be weaker – so you are continuing to act like animals. It is true that you have evolved from lower animal life; you are nothing more than slightly advanced monkeys since you hold 99 per cent of their DNA. You made a series of movies called Planet of the Apes, a story about barbaric apes who control your future world. For those of us who monitor your world, the same scenario has lasted longer than a few movies. The difference is that *you* are the ruling barbaric race.

Despite your primitive mindset, you are starting to wake up. Your evolution has become greatly accelerated when compared to the normal and painstakingly slow pace of evolution in its natural form. Your higher evolutionary pace is what separates you from the lower animals. It has separated you from nature, thereby throwing it out of balance due to your break from its cycles. It is therefore essential that you develop the wisdom and skills to prevent a complete breakdown of the natural world. So far, you have not done a good job of it.

That is why I am here, offering this book of lessons. You need guidance. You are children who have stepped out of Eden, out of the natural cycles, to build a plastic world that you attempt to control with warped minds – minds that operate so far out of balance that you fail to see the danger of your predicament. Guidance is the rule of the day should you be able to cast your egos aside, even momentarily, to listen and heed these words.

You are caught between two worlds. From a genetic standpoint, you are hybrids – half animals and half gods. One of your great philosophers, Plotinus, once said, "Man is poised halfway between the gods and the beasts."

Plotinus was a very wise man. The *gods*, as you referred to them, were real. They were not myths, but *became* myths. They were at one time on your planet and interbred with natural man. This is found in many of your holy books, including the Bible.

> "the sons of God saw the daughters of men that they were fair; and they took them wives of all which they chose."   (GEN 6:2)

> "There were giants in the earth in those days; and also after that, when the sons of God came in unto the daughters of men, and they bare children to them, the same became mighty men which were of old, men of renown." (Gen 6:4)

Similar stories, all relating these same events, are found in the mythologies and religions from around your world. There is no denying this fact if you look into it deeply enough. With godly genes being incorporated into mankind, it became our obligation to look after you. In a certain sense, you are our offspring. We honor your free will as we honor our own, so no longer attempt to directly interfere with your path. Only at the brink of disaster would we ever step in and do so. Although I have always favored it, I cannot guarantee that we would help you even then. That day may one day come, but before it does, I am here to nudge you (not judge you), with guidance, so you may be able to change your course with wisdom. You must help yourselves and not depend on others. You must learn the needed lessons for yourselves and perform the actions that prove your accomplished wisdom. Allowing you to do this is what mutual respect is all about. You must therefore extend this same respect to your own brothers and sisters across the globe, and begin to work *together* on your problems, not as enemies.

You currently cause more unnecessary problems and heartache among yourselves than any known species in this galaxy. Only when you prove that you can work in harmony among yourselves will the invitation be extended for you to work with others who are from elsewhere. You have logically deduced that you are not alone in the universe, but have not officially been made aware of this fact among all your inhabitants because you are simply not ready. No intelligent species from elsewhere will reveal themselves to you, although you may eventually discover them. To put it bluntly, no one from an advanced world wishes to embrace you in any way because you would clearly not extend to them any mutual respect. You do not do it among yourselves, so you will obviously not employ it elsewhere. The only way you would ever respect a higher race from elsewhere is if you were forced to fear them. But higher races are not interested in spreading fear, so they prefer to keep their distance from you – at least for now.

You have become an interesting "experiment" of ours, and it is often debated as to whether or not it will succeed. In fact, there have been times when earlier versions of the experiment had to be abandoned, and you were forced to start all over again, with a few survivors on a barren, devastated world. That time may come again, brought on by your own shortcomings combined with powerful natural events.

The mutual respect that you need to cultivate must extend to all living beings, not just with other races of mankind or other different types of humans. All life is sacred, but you treat it with little respect. You slaughter millions of animals every day and think nothing of it. It is "business as usual." Although this is done primarily for food, it is done without compassion or respect for these smaller lives. You have become not just food suppliers in this respect, but merchants of death and suffering. You may be smarter than these beings, but *you are not better than them*. Life is life; it is all sacred, every part of it, and it must be respected. You have grown supremely arrogant and mistakenly believe that you are 100 per cent gods, and can do as you wish, but you still act like animals. You are a savage contradiction that refuses to look in the mirror.

You are a hybrid race, half god and half savage, and this has caused great confusion among you. No other species in the animal world can match your power and influence over the world. Yet no other species in the world suffers with as much anxiety and mental illness as experienced by humankind. We watch over you and try to guide you, but cannot intervene directly. This is your path to follow. You must work through your problems and elevate yourselves to another level – otherwise you will not survive. It cannot be done for you. You must do it for yourselves. Without mutual respect, you will, however, fail. It is time to put your differences aside and begin working together to save your place on the planet. If you continue to exploit each other through the illusion of "superiority," you will all suffer and lose.

You know that mutual respect is something that needs to be worked on in the areas of race and religion. Your awareness has been directed to work in these areas, so you have made some progress here. But the most important part of this message, on the subject of mutual respect, is directed to those who operate at the higher levels of economic power. These individuals and groups of individuals are often hidden from public scrutiny, their names almost never appear in your media, and they operate with almost complete impunity. They work in a vast Machiavellian network that answers only to itself, to the point where entire world governments must bend to its will, whether they be in opposition or not. With such power, it is easy to view the masses in third world countries as "useless eaters," as they are sometimes called. By doing so, they are treating their fellow man as cattle, and denying the respect that each and every one of you deserves. Therefore, you can make all the progress in the world learning to respect other races and religions, but if those who control your world's purse strings learn nothing, you will all remain as slaves.

Greed, wealth and comfort breeds disrespect toward those who have less. The average people in society are forced to use the same character traits that have been set in place in order to preserve themselves in a world controlled economically by a "might makes right," survival of the fittest mindset. Those at the height of economic power are so drunk with their power that it supercedes the power of the spirit, which waits in the wings for your awakening. An atheistic attitude is generally employed because it allows conduct without moral values and therefore, without conscience. The idea of answering to a higher power is ignored because they consider themselves to be that higher power. They have sold their souls – because they act without them.

The average person *who has awakened* does not need to do this. If each person is willing to surrender a small degree of comfort in order to improve the lives of others, society itself would be considered awakened. You would not yet be enlightened – you still have a ways to go before having this realization – you would simply be awakened. You cannot become enlightened beings until you are first *awake*. What you must learn is that it is far better to live in an awakened world than one based on exploitation, and that it is worth working for. You must start to make the needed changes – changes not needed for you, personally, but needed for the benefit of mankind. There is a distinction to be made here. Your economic leaders are trying to shift the paradigm while still holding their power, and that is not possible. They act in your stead, so you are complicit in allowing this path to continue. You are trying to trade the use of things like big oil into other forms of renewable energy, while funneling it through the same flawed power structure. If the apples continue to rot in the barrel, new apples will not solve the problem. You need a new barrel.

Exploitation must always occur, at least on some level, for necessary exchanges to occur. But what you are failing to do is to properly compensate for the exchange. You instead try to leave the victim with nothing, or with something less than fair. You willingly cheat those whom you should be loving and helping. Currently, the ultimate goal of exploiters is to be so crafty that the victim will never know he has been cheated, or will discover the fact much later. Sometimes, agreements are signed that have hidden provisions in small type that the signer would otherwise never agree to. The proper goal, however, should be to fully inform the intended "victim" of exactly what is going on in plain English (or whatever language is used), and offer something tangible in return, that you know they would be happy with. If you cannot afford to make a "win-win" offer, then no offer should be made and no action should be taken.

Humankind is more interconnected than you know. You all need each other to get by, but you instead believe that you need to *exploit* each other to get by. Less for others does *not* mean more for you. It may work that way temporarily, but everyone eventually loses when exploitation runs rampant. However, mutual respect among you will insure your eventual success. If those in power would embrace this idea it would be somewhat of a miracle. They cling to their power like a spider, sucking on the juices of its cocooned and unconscious victim. They have the masses "wrapped up" in this very same way, due to years of consolidating their power and forming structural alliances that, if dismantled, would bring down the current paradigm.

The paradigm, however, must fall. None last forever. You are stretching this one out, squeezing every last bit of energy out of it – but that only extends the pain and discomfort for everyone at the bottom, and will make it worse for everyone once it reaches its final stages. It can no longer operate with much strength no matter what you attempt to do because it has run its course. The "lesson" is about to be learned. All structures and alliances based on exploitation do not last very long, despite the clever and desperate maneuverings that you are witnessing. The system gets more and more sophisticated in order to maintain itself, but the weight of it – it's sheer mass – has grown too cumbersome to remain efficient and operational. Your geniuses will run out of answers. Simplicity, fueled by trust and mutual respect, will be your next move. It will soon be your only option. Make your plans, and open your hearts.

Exercise

Walk in My Shoes
   You need one day to do this exercise. Set it aside ahead of time, make no plans or cancel any minor ones that exist. You are going on a mission – but only if you feel that you would not be in danger. This is not a required exercise, but recommended. The decision to do it is yours alone. Therefore, you are entirely responsible for your actions and their results. If you have already done it once and are repeating the exercises, it is not necessary to do it again.
   Men can do this alone; women should go with at least one male in each case (not with another woman), or spend the day in the safety of a shelter, should there be room. There is a Native American saying that says, "Do not judge me unless you walk a mile in my moccasins." From there, it transformed into the white man's version of "walk a mile in my shoes." Which you may decide to do.
   You are to become homeless for one day. Research ahead of time and find out where the homeless in your area congregate or seek help, and if you would be welcome in these areas. Do not dress well, leave early in the morning and spend the early hours alone, getting accustomed to the loneliness and isolation. If you believe your safety to be an issue, then spend the entire day away from the homeless. You are to bring no food or money with you. At all. When you get hungry you must figure out how to get food. If you fail to get food and go without, consider it a day of "fasting."
   Should you meet homeless people that you deem to be safe, get to know them. The best course of action is to go about your business and not volunteer any information that could anger anyone. If only to avoid hostility toward you, have a hard luck "cover story" ready. Otherwise, be truthful if confronted. If you are honest about your life, do not put on airs, or try to appear "better" than anyone. Be smart. Offering the truth could cause resentment, or it could gain you respect. If you are feeling

unsure, let them know that it is too early to share your story and you will tell them later.

If you need money for food, beg for it – but never in front of those who would truly need what you gather. Going to a place where friends or relatives might see you is probably not advised – although they would certainly not be looking for you, and a "costume," of sorts, would help. If you have musical ability, bring your instrument and put a hat or container out for donations. Do this only if you feel safe enough to do so, as the instrument may have some value and you could therefore be jeopardizing its loss.

You may think that you do not really need to do this exercise, as the lesson that needs to be learned can be conceptualized to a suitable extent. You are wrong. Nothing could replace this experience. You can cut it short and go home as night begins to fall, especially if you are female, but a full 24 hours is recommended depending on the sleeping arrangements that can be found. Do not endanger yourself. A shelter is not advised if you would end up taking a bed from someone who would truly need it.

It has been said that many of you are only one paycheck away from being homeless. If you do not get paid just one time and all your bills become delinquent, it can be impossible for some people to ever catch back up. One health problem without having insurance could devastate you or your entire family, and you could find yourself out on the street with practically nothing. There are many of you who can attest to this fact.

People often look down on the homeless or simply try to ignore them. A number of them are mentally ill, so it is advised to steer clear of those individuals because, as unfortunate as their plights may be, they can be unpredictable. Many, however, are just like you.

Twenty-four hours or less is not enough time for you to understand the true effects of homelessness, but it will provide you with a taste. You will be able to go home to proper shelter, clean yourself from what little hygiene was affected and not be forced to eat questionable food out of desperation (or out of trash cans). The main purpose is for you develop mutual respect for those who are like you, but who experienced some bad luck. If you can increase your mutual respect for such people, then this exercise will reap you greater rewards than you are currently aware of.

Lesson Seven

Wealth and Success

This lesson will create a new value system for you and completely redefine what true wealth and success really is, based on a shift in your perceptions. You should by now be able to use your newfound focus and clarity to help in this redefinition.

One of the most damaging mistakes you have made has been to equate the accomplishments of wealth and success in a synonymous way. To believe that wealth equals success, or vice-versa, and to act in ways that pursue wealth as your primary goal, often leads you in directions that are self-destructive. Many of your people have died without being truly successful at all, because you pursued wealth *instead of* success. Wealth and money are false gods, pursued by blind followers who know nothing of their own souls or inner spirits.

Each of you is here for a specific purpose. No one else can do what you do. By looking deep within yourselves, you can uncover your true purpose. The idea of pursuing wealth first, instead of your real path, is a diversion. You often think, "If I become wealthy first, I will then have the time and freedom to 'find myself.'" Nothing could be further from the truth. You will never find wealth without knowing who is looking for it.

Your thirst for wealth, therefore, is never quenched. No matter how much you get, you always want more. Always. It is an addiction. There is nothing evil about having money. It is what you do in your quest to achieve it that is often evil. The level of money one achieves is often in direct proportion to the amount of greed one harbors within their soul. At the same time, those who have noticed this fact have turned into the most generous benefactors the world has ever seen while trying to correct this flaw and help others at the same time. You need more of such people. When you look at history and examine it closely, it becomes clear that many people who become *excessively* wealthy do so by exploiting others or doing damage to their fellow man in some way. When actions are fueled by greed, you will blindly trample over each other in your quest for wealth. Terrible things can happen – and often do. Material things become more important than relationships, but without nurturing relationships, you have *nothing*. All you really have is each other. You could have all the wealth in the world, but if you have no one to interact with in a truly meaningful and emotional way, you have nothing. It sometimes takes people a lifetime to learn this lesson – so I am sparing you the expense and possible heartache by telling you now. Of course, many of you will fail to listen and will learn the hard way, but for those who truly know how to listen, take note. You have just been blessed.

How you spend your wealth, in addition to how you achieve it, can be just as destructive. Power results once you have achieved a certain level of wealth and it can be easily misused. Those who lack wisdom and maturity are the biggest offenders.

You must learn who you are first. By doing so, you learn your proper path. Your proper path will never fail you, or allow you to exploit or damage others, including yourself. If you happen to harm others for your own personal gain, then you are not on your path of purpose. You cannot write off the victimization of others as being "the law of survival," or justify it with "that's just the way it goes, for some to win, others must lose." Discovering who you really are brings forth the powerful realization that you are all deeply connected in a way that goes far beyond a cute, hopeful metaphysical mantra. It is a reality. When you see the big picture and learn exactly where you fit into this important, interdependent tapestry, then you will no longer be victimizing or exploiting anyone, for any reason (unless it is totally and completely accidental).

Wealth and success can destroy you just as easily as it can harm others. You have seen many famous people self-destruct because, despite being famous, they never truly found themselves. The list of famous celebrities who self-destructed goes on and on – people who were admired and looked up to by millions. But when their lives are examined more closely, it is discovered that they were troubled souls in some way, who hid their addictions or flaws from the public and battled their personal demons from behind closed doors.

These "demons" will kill you. If you internalize it, that is what happens. If you externalize it, you kill other people instead of yourselves. Externalized power most often results in war or other forms of violence. Humanity's preferred means of flexing its power is through warfare, waged in order to dominate one another and lay claim to the material possessions of other neighboring entities. You are still brutish and nasty animals. There is no escaping this fact, no matter how much you wish to candy-coat it through your displays of "patriotism." You are slowly learning that there are no real winners in war, and that it brings failure to everyone involved, in one way or the other, as a consequence.

Wealth and success must be achieved in other ways. For now, however, creating war and supplying weapons (sometimes circuitively to both sides) in relatively backward countries, filled with ignorant people, is the perfect money making industry. You are masters at keeping enemies antagonized toward each other so you can sell them billions of dollars worth of weapons. Your merchants of death finance and pay for the carnage, but will never be found anywhere near the trouble. They see themselves as being "better" than those below them who have no access to such wealth or power, but they are in fact the most cowardly hoodlums of them all, despite their opulent lifestyles.

It is true that brutally beating, or even killing, an opponent makes you look much better by comparison. The victim either never recovers from the loss or does so very slower; while the victor basks in the joys of victory for just a short time before wanting to be sated again. The victor grows tired of the spoils more quickly than a loser heals, and the thought never enters his mind that a win-win situation could ever be possible. In private, the ego wishes to share nothing, not ever – but sometimes, in public, it will extend some deserved credit to maintain a positive public image. The ego is a vicious, conniving "partner" that revolves in a closed universe around itself. It must keep attacking others or can sometimes turn on itself and cause you to self-destruct.

Your lesson in this section is to learn how to channel your energies into more positive directions, without listening to the ego. It constantly chatters in many of your heads, and you must learn to quiet it – or at least pacify it enough to allow deeper and more meaningful thoughts to surface within you. Your ego deceives you into believing you are on your right path, but your true path is rarely found by listening to it. The ego is like the outer skin of an onion, and can be just as flaky. You must peel the papery veil away, and get to the meat and sustenance below it. Many of you are completely ignorant of your deeper self; completely unaware that it exists. You have never encountered it, nor do you want to.

The vast majority of you allow your immediate desires to run your life. You therefore jump from one desire to the next, allowing them to "play you" like a violin. Scattered "notes" get played here and there by the ego, but the real tune that defines you never gets played. You have less consistency than those who are grounded in the higher Self and can receive guidance from the depth of their souls. Your life can be a major symphony and a true work of art, but you spend it dancing, instead, to the tune of your "monkey mind," as it chatters its demands to you like a spoiled child. Being a slave to a spoiled child will never make you wealthy. It will, however, steal your true wealth away.

True wealth and true success comes to those grounded on their path. Such people hold no doubts whatsoever, as to what they are here for. They have identified their mission in life, so their wealth and success are measured by their deeds and the satisfaction they generate by performing them. Although they are human like the rest of you, greed is a far less substantial factor in motivating them. Higher ideals and moral principles guide them, and they are not distracted like a donkey running blindly after a carrot tied to a stick.

It is time for you to find your guidance, and throw out the interloping ego that steals your lunch every day. You are no longer in grade school, and do not have to put up with this anymore.

Exercise

True Wealth

Take five minutes a day and detach yourself from your ego. The best way to accomplish this is to watch your actions as an observer rather than a participant. Pay attention to your every action during this time and study why "this person" (meaning you) is doing it. Seriously dwell on exactly what is happening. Ask yourself, what real purpose is being worked out or accomplished from the actions being taken?

You will find it difficult to remain an observer for very long – it is too uncomfortable for the ego to have a higher part of you looking down over its shoulder. The ego likes to be in charge and cannot stand close scrutiny for very long. The ego wants its demands on a constant basis, like a fat, chattering baby bird demanding constant food from its mother, and does not want you questioning why it wants these things from you. It feeds off its desire for instant gratification, which is often not in your better interests. You must begin to introduce the idea of the ego surrendering, or at least sharing, its power. Your "spoiled child" would then have to "grow up" – but it would rather die, kicking and screaming, than to give up any of its power. So during this exercise the ego will jump back in control of your thoughts before too long. Your job is to gently push it aside and begin observing yourself again each time this happens.

If you can maintain your observations for just five minutes at a time then you will have mastered your lower consciousness successfully. It will be enough to make an impact; enough to get the ego's attention. If you can do this every day for a week you will cause the ego to shift ever so slightly toward an acknowledgment of something higher. And if you can do it for one month, you will begin to see a shift in your entire, everyday consciousness. Your mind power will improve dramatically because you will have invited your higher consciousness into your normal, everyday lower consciousness and if it accepts it enough and *integrates it* for up to five minutes per day, then parts of your higher consciousness will slip in and help you more often. Your higher Self, with its great wisdom, always sees the folly that your lower self engages in – but the difference is that you will become conscious of it. And your lower, natural mind will be acclimated to a very powerful ally that will, at first, not be very welcome. But with persistent practice, the two minds may meld together and begin acting in unison on certain problems, even if on a subconscious level, to accomplish great things. Your intuition will become slightly stronger. If you can listen to it and act on the advice of this very quiet inner voice, then you will begin to engage yourself in fewer trivial matters and diversions. You will start devoting yourself to things that are more meaningful.

The real challenge, however, that most people fail to meet, is in creating this unity every single day for a solid five minutes each time. This exercise sounds easy, but is extremely difficult. Five minutes is a longer time than you think, and many things will happen to thrust you back into normal consciousness. But you must gently go back and start over again each time, and do not quit until you have maintained the observation for a full five minutes without losing it. You will reap few benefits unless you can accomplish this every day for at least a full month. This is an exercise that can expand your consciousness and increase your mind power if you can maintain enough focus to do it.

This exercise is important because it will move you further toward a higher consciousness just enough, as a first step, while not being designed to keep you there. Consciously remaining in this state is a challenge and would only become easy or "natural" as a direct result of your *kundalini* being raised. Kundalini is a powerful energy hidden within your bodies that few of you ever access. This will be covered in more detail in Lesson Eleven. Lesson Eleven will be of less use to you, however, if you cannot maintain this current exercise for a month.

With practice, this entire lesson will open a spiritual "valve" within you and start an energy flow that will direct you to a higher level. When you begin operating on this higher level, your values will begin to change. A new perspective on life, including who you really are and why you are here, will continue to unfold.

PART THREE

ADVANCED LESSONS

Lesson Eight

Understanding Yourself

This lesson will enable you to identify what kind of person you are and, using this information, get to "know thyself," as the ancients would often do.

The ancient Greeks and other cultures from the past had an appreciation for inner development. The great mystery schools were concerned with the perfection of man. Today, humankind has regressed. You are now more concerned with the perfection of *things*, rather than yourselves. And you call it "progress." Those cultures that lived in harmony with nature and who lived more simply, never experienced the levels of anxiety and depression found among you today. For proof of this fact, you should travel to a remote part of today's world and visit an isolated, primitive tribe still living simply today, without the trappings and "conveniences" of the modern world and see how truly happy they are. They are in tune with creation and live as part of the natural world. The most unhappiness they experience comes from the encroachment from the outside world (from your material-based consumerism), which threatens their way of life.

You are not advised to match your lifestyle to theirs' or throw out your past life and move in with these people. It is too late for that because you were brought up differently. You have a "set pattern" of existence that has been established – but from within it you can still make important changes in your life. You can still reach an inner part of you and try to perfect it, rather than reaching out for "things."

Before I can guide you in an understanding of who and what you really are, it must be determined where you are in your own personal development. In general, there are three types of people or levels of awareness. Knowing yours will provide you with a starting point.

Type One

The first type of person is one who lives as their body. You identify yourself as the body. You exist for its immediate needs and always satisfy the body first. Your mind, however, is the key. It is your mind, or level of consciousness, that is connected so strongly to the body that all that your mind really wants to recognize or identify with is the body and the materialistic pursuits that it needs. Your focus is primarily an exterior one. You are totally and completely trapped, without knowing that you can move beyond this because, in most cases, if you knew you could move beyond it, you would. Just a few of you at this stage sense there is something beyond this and that there is a new way to think, and ultimately to act, but are too afraid to embrace it.

If you mostly identify yourself with this state of being, that is fine. There is nothing wrong with you. In fact, this is exactly where most people are and you can consider yourself "normal." If you wish to learn more about your true self and feel comfortable doing so, then the following exercise may help you, no matter what level of awareness you are at. Previous lessons in this book always worked up to the exercise appearing at the end, but in this lesson it is best for you to encounter it here.

Exercise

Touching the Past

Upon completing this exercise you must take some time to digest what has happened. You will need time to determine the extent of its impact upon you. Do not, under and circumstances go immediately to the next exercise that follows. You must do this exercise every day for at least one week to get the benefits from it. When the week is over and these seven sessions are done, you can move on to any of the exercises that follow this one. It is fine to do this exercise at the same time you are carrying on with your 30-day period of self-observation as requested from you at the end of the last lesson – but you should do them at different times in the day. This point in the book is the most challenging in doing the exercises, but meet the challenge. This is a key moment. Do not "drop off" and just do the reading. You must engage yourself with the material to reap its great benefits.

Even if you consider yourself not to be part of this Type One group of people, it is a good idea to start here, with this exercise, and carry this through as a "process." It will only enhance the results that you will ultimately have. Pick a good time of day to do this exercise so it will be easier to stick to the plan and complete it every day.

Sit down, take some time and become very quiet inside. This is essential before you can continue. Get as relaxed as you can and take a number of very deep breaths. Before you close your eyes, read the following instructions so you will know what to do.

You carry within you all of events in your past. So within your mind, go to a place in your past that gave you incredible peace. Visualize it. Take a few minutes, once you establish yourself there, to relive it. Remember the smells, and experience that blissful and relaxing time once again, completely and fully. You should become fully relaxed, being back in this special place. Spend as much time there as you feel necessary and remember every detail and feeling. Smells or almost forgotten feelings may resurface. You will be there in your mind, but not with your body. Do it now, and after you do it, come back to the book.

Upon your return and with your eyes open, spend some time reflecting on this visit to the past. Ground yourself back in your body, back in your normal state of being, but make sure to stay relaxed. Maintain this peaceful feeling. Once you feel ready then, with eyes closed, ask yourself the following three questions, and dwell on them deeply. You should do them one at a time, instead of together. Before you ask them, make sure that you are still deeply relaxed.

Seek deeply within yourself for the answers because *they exist within your past*, back before you came here, back before you were born. Do each question one at a time rather than together. Once you have explored the possibilities and dwelled upon them, go to the next question. Remember, having an answer is not important right now. Exploring the questions is far more important. Spend some time dwelling on each one. Examine it from a number of different approaches. Come back to the book after a few minutes of introspection for each one.

1) Who was it that just experienced that?

It is important to think deeply about this next question as well, since it refers to "what" rather than "who." It involves something more general than the previous question.

2) What am I really?

You may have felt what you are, but you may not have been able to explain it fully, or in a rational way using words. Despite humanity's tremendous dependence on words, the best way to explain yourself, you will realize, is through your actions and their symbolic "triggers." The great mythological tales and psychological archetypes are all tied in more strongly to actions than with words. Dwelling on this question can be a very powerful experience if you can go deeply enough to discover the hidden keys. It has the potential to create a major breakthrough in your life.

Your third and final question requires an even deeper plunge beyond, because you must ask yourself this question very, very *slowly* within your mind. The one mechanism that has always defined "you" is your mind. Whatever does your thinking – meaning whatever creates your thought impulses and gives them meaning – even before your emotions enter into the equation, is what defines you. But the mind always jumps around very quickly and is easily distracted. So you must slow down the very thought process itself and examine the *impulse* more closely when you close your eyes and (do it now) *slowly* ask the following question, a number of times. Take your time, and only finish when you feel comfortable – not when you receive an "answer," but when you feel comfortable.

3) Who is the Thinker?

The exercise is now over. As you continue practicing it over the coming days, or even months, you must determine, through deep examination, if your mind creates the impulse of thought or if it merely receives it. Is the ego your thinker? Or is the thinker your soul? Or are your thoughts the result of natural chemical and electrical processes in the brain? Each of these answers to question number three will make equal sense until you *experience* the answer. The answer will remain useless and nonsensical to you unless you can experience it in some way. But if you are steadfast in your pursuit of the mystery that is "you," it will eventually come forth and reveal itself. Many of you will not remain steadfast and when the answer fails to surface you will continue to live in darkness. For those who succeed, it will be a day that you will never forget. It will not mean the exercise is over and you do not need to return to it. It means even more so, that you should return to it periodically. Once the connection is made and continually practiced, you can put yourself into a place of power *just by asking the question.*

The important thing for now is that you have disengaged the ego through this exercise. When the mind and body are tied together so closely and focused only on outer things, it creates such a narrow view that only an individualistic ego functions. It appears as the controlling factor in everything you do (except when you are asleep). Materialism then rules your life. Despite the puffed up feeling of importance that the ego provides, you will realize through this exercise that a tremendous freedom can result from leaving the ego behind. You are *not* your body and therefore, the ego, with its strong attachment to it, is a *self-made illusion.* This first level, Type One, creates a degree of insecurity because you know very little with any certainty beyond the material. The ego is a reflection and ultimate expression of your own insecurity, and creates a pale substitute for true spiritual power.

With this exercise, however, you should have experienced at least a small flash of your true spiritual power, and touched on the revelation that you can function, truly function, while aligned with your higher Self. This connection may not be maintained for very long each time you do this exercise, but if you are consistent in this practice, you will begin acting in ways that reflect who you really are, more and more and every day. Your path will open up in front you, and you will no longer be forced to hack at the branches of a clinging ego to see your way through – you will instead find a clear path in front of you, whatever it may be, that makes perfect sense for you to follow. You will literally see it open up to you as you progress.

Do this exercise every morning or every evening for at least a week, preferably a month. Do not miss a day! If you do this, a powerful shift in your thinking, *and your actions*, will eventually occur. It will happen. As it unfolds, no one should have the right to fault you about what you are doing and what new changes you are making in your life – but they will. Others have a conception about what you should be doing to fit their "reality tunnel" of the world and how you should fit into it, but the only reality you must pay attention to is your own. If you can change your perceptions, it will change others' perceptions of you as a result and many will not like it. You are not here to dance to any one else's tune, despite how much they may want to manipulate you. You all manipulate and exploit each other, but your mark in the world will leave more of an impact if you can minimize the amount of manipulation that you must fall victim to.

Your power and influence in the world will be greatly enhanced once you gain a better understanding of who you are and what you are doing here – *really* doing here. You no longer have time to waste. Let those who remain as sleepwalkers waste their time, because you are now awake – or at least working toward it.

Type Two

The second type of person is one who lives *in* the body, as opposed to *as* the body. This should be the minimum result for all who complete the previous exercise with focus and diligence. The realization has occurred that you are more than just the body and this something "more" has far more to do with your true identity than your body ever could. Your body is just a shell and acts as an instrument for the work you were sent here to do. When the body wears out or is destroyed you get to move on, taking with you the new things you have learned.

Your soul has energy. Scientists have told you that energy can neither be created (from nothing) or be completely destroyed. It can only be manifested into a different form. So when you are done with this body, you will move on. The riddle of the afterlife is not a riddle at all. The answer is quite simple, as just expressed, and comes from basic scientific principles. You are energy, and will simply take on another form when the body stops working. As many of you say, "it is not rocket science." You treat it that way, however, because so many of you are frightened out of your minds at the thought of physical annihilation. It is your ego that is fearful, because it is your ego that will die with the body – not you. You are not the fearful little ego that has convinced so many of you that it is your actual identity.

You have traded in your true identities for inferior counterfeit versions that run most of your lives – usually in trivial circles. If you moved beyond this and found your true selves then not only you, as individuals, would change for the better, but the entire world would transform itself overnight. That is the purpose of this book... to move you closer to this transformation and the discovery of your true selves. So let us move to the next type of person.

Type Three

The third type of person is very rare indeed, because he or she lives *through* the body, rather than in it, or as it. Becoming this type of person should never be a "goal" of yours. If you become aligned with your Self and are on your proper path, then you will carry out your expressions in the ways that are needed. That is more important for virtually all of you. Your life path will be on track and you should never strive for things beyond your reach. You will become a Type Three person if and when you ready, and no earlier than this. You cannot force this to happen. With that said and understood, we can move forward.

You, meaning your soul, is from the spirit world. You have descended here to engage in the lower material dimension because this is where you can advance. The challenges you encounter in the physical world are more rewarding than what is found in an existence that has less density. In the physical world there are solid obstacles that must be overcome, and the actions of others pose endless challenges for the advancement of your moral character.

You have a higher Self that observes this life. It benefits from your better deeds and attempts to guide you, through your intuition. It is your Overseer. Although I am the Overseer for all of mankind, you have your own personal Overseer in the form of your higher Self. Use it. Cultivate it. Most of you are cut off from your intuition or ignore it. Type Three people are in full touch with their intuition and it never lies to them or lets them down. The ego, however, does this constantly and leads you down false paths. Your ego gets you in trouble all the time because it lacks the wisdom of the higher Self. The exercise at the end of the previous lesson (Lesson Seven) is good because it allows you to touch on a spiritual form of consciousness, however briefly.

Know Thyself

Know thyself. This is sage advice attributed to your greatest philosophers – Pythagorus, Heraclitus and Socrates included. But it originally appeared after a visit of mine, found inscribed on the Temple of Apollo at Delphi. My verdict as to how well you have done with the advice is failure. You started out well for the first century or so, but it fell quickly into misuse. Instead of putting it into practice on a continuous basis, which takes *effort*, you found it better to ascribe it to a host of men wiser than you so you could comfortably stay away from ever pursuing it. The adage *know thyself* has followed you down through the ages like a ghost floating through the mist, ever so elusive from your half-hearted attempts to grasp it. Today it has lost almost all of its meaning and has grown into nothing more than an insignificant catch phrase. Entire schools were based on it, but you have diverted your power and focus away from it, toward outer pursuits that mean virtually nothing by comparison. I have tried to guide you, but you are blind and spoiled children still struggling toward an awakening, often kicking and screaming on your reluctant path toward enlightenment. So now, instead of inscribing short guidelines and affirmations on temple walls, I have determined that you need an entire book in order to learn about yourselves and how to act. You are on a downward spiral and need to reverse your direction. This book can help, *if you apply yourselves.*

Those who give "expert advice" in the modern world are often those who are furthest away from knowing themselves. You are basing many of your daily decisions on the advice of fools – trusting your most important decisions on the words of self-professed "experts" who do not know you, but claim they know what is best for you while, at the same time, are reaching into your pockets or purses for your wallets. Even those who know you best often give you bad advice. So why would you listen to these further removed "experts?" Your media have trained you to trust them, but it is a sad diversion. It not only depletes your wallets, it robs your spirits.

The key to your future is found in first balancing and then listening to your innermost self. Society does not like it when people make individual and unique decisions. The large corporate entities that steer the economic machine want you to be predictable; they want you to do as you are told and to buy what you are told. They want you to listen to *them* and respond in material-based, outer ways rather than delve into any form of inner exploration. Money has become your God and they, in large part, control the money. Therefore, they expect to be listened to as though they were gods, and it is your job to *obey them.*

However, there is one thing that they have forgotten. And that is, that the kingdom of God is within you. Social engineers, no matter how clever they might be, can never control individuals who are aware of this fact. Money has no soul, but you do. The kingdom of God is within you. When you are aware of an entire kingdom at your command and use it, there is less of a response to mass marketing tactics. A true kingdom resides within your very self. This is not just a Biblical figure of speech. You have, at some point in your life, experienced this power within you. Think back and remember what happened in your life when a tremendous power surged up from within you and you triumphed in an amazing way. You have all had these special moments. What was it that you did that amazed yourself and others? What was that feeling like when you completed the event, or task or journey? Chances are, you listened to *no one* when this happened. You followed your heart and somehow knew deep down, that you could do it – almost always in *direct opposition* to what others were telling you.

So why would you listen to others on a daily basis, who are quite likely blocking you from similar achievements? You are capable of amazing things. It is time to get in touch once again with what truly drives you, to assess what you are here to do, and to get busy doing it. Your path awaits you.

There is a mystical belief from the East in your world that claims physical reality is an illusion. They call it "maya." There truly is an illusion involved in your world, but it is not the world itself that is the illusion. The world is real. It is a solid, physical thing and you are in it, here and now, experiencing it. The illusion is found in your *perception* of the world. You view it in such a warped way that you believe you can force it into behaving in the way you demand. Sometimes you can do this in limited ways but the flow of life – the main current that carries everything together – does not bend to the whims of one of its living parts. You grab and manipulate the physical world to your designs, but fail to see the *flow*, and how you are just a part of it. You will not nor will you ever conquer the flow, or the Tao – it is impossible. Your consciousness is deranged and has created this illusion.

There are many different levels of consciousness. The level that most of you operate with in daily life is amazingly inefficient, despite being considered "normal." This does not mean that I advocate the use of anything outside of the mind that can be used to change your consciousness, like mind-expanding drugs. Consciousness is needed for survival, and survival is about awareness. Drugs or mind-altering substances of any kind remove a certain amount of awareness of the surrounding world from you – a world that contains vital information that you need to be aware of, so this is not recommended. You need to gain more awareness of your surroundings in this time of crisis, not escape your surroundings or diminish your awareness to any degree. In general, the goal is to sharpen your awareness, not to numb it.

Some of you do in fact need medications that affect your consciousness. They are usually taken to prevent your mind from entering into dangerous or delusional states, which is of course necessary. If the mind has chemically or electronically malfunctioned to the point where it decreases your awareness of the surrounding world by itself, without the use of drugs, then certain drugs may be needed to correct the imbalance. My point is that you must establish a balance before you can begin working to achieve an understanding of who you are. There are no shortcuts in the form of pills or substances that will achieve this understanding in any prolonged or meaningful way.

Artificially changing your consciousness might pull back the veil temporarily and provide a glimpse of your true self, but the glimpse will never be maintained without lengthy practice and skillful control.

Another key to your innermost self is meditation. Lengthy practice is often required. If done correctly over time, meditation will create a balance of the polar opposites within you. You cannot tap into and receive authentic guidance from the higher Self without first achieving an inner balance. Otherwise, you may get only fragments of higher Self-guidance, due to interruptions from your conscious mind. More instruction will be found on meditation in your Lesson Eleven exercise.

In addition, people who listen to inner guidance and who are not properly balanced are often listening to parts of the subconscious mind that have issues or unresolved problems. These problems get buried deep within the mind earlier in life through trauma and release themselves later through stresses or other triggers. Voices often result, *heard or unheard*, that lead you terribly astray. Strange or bizarre compulsions, acts of violence or criminal activity can result. Many people do not know why they engage in harmful or compulsive acts and they do not know how to stop. It is like they are possessed. It is not the devil at work; nor is it God, as some may believe. It is the result of unresolved issues that, if brought to the surface through therapy, deep inner work, or meditative practices, can be resolved. If balance is restored then a more direct path to the higher Self is created.

When this happens, all criminal activities or bizarre compulsions will cease. You will then begin to know your true self rather than a counterfeit one. For example, many priests once engaged in self-flagellation (and some still do) – the practice of painfully whipping oneself in "service" of God. This is the result not knowing yourself; you are instead attacking it. You cannot serve God by attacking yourself. Your body is the temple of the Lord, as is often expressed, so it makes little sense to injure it. In reality you are all part of God, so there is little reason to attack Him. What you need is balance.

This lesson has provided you with a deeper understanding of who and what you really are. With this knowledge brought down into your normal consciousness you have at least begun to explore ways that will be comfortable for you to work with it. You will soon know the power of the great ancient goal to know thyself, by following the path set forth in this lesson.

Lesson Nine

Understanding God

Humankind does not understand God. I am here to help you do this because if you truly understood God you would act in accordance with the highest precepts of your religions. Those masters who founded your great religions understood, at least in part, what God is. But humanity as a whole has watered down and polluted its religions so that the original teachings cannot be followed properly. You have lost your path to God. You cannot find this path again without knowing where you need to go. When you understand God, you will know the path again. That is the purpose of this lesson.

You most always interpret your religious teachings in outer ways. You look outside of yourselves for the answers – for a God from heaven or angels who can come down and guide you – yet the answers must be searched for deep within yourselves. You are chasing material things and worshipping material things that will never result in salvation. Your crosses, flags, statues, holy texts and symbolic icons are illusions and when these illusions do not quite work you simply create another one in its place. It never dawns on you to *look within*.

God is within you. All things are connected through God, and you are connected to God from within. But you continue to focus "out there." You fight your wars and kill each other in the name of God, focusing outside of yourselves. When two men love the same woman they will often try to kill each other. That is what you are doing with God. Yet, when two men love God it is meant to unite them, not pit them against each other. But you must first *find God* before you can truly love Him and unite yourselves.

Each of your religions separately claims to be the only path to God. They proclaim all other paths to be trodden by heathens who will never enter into the true heaven because you – or rather, your religion – "owns" it. Those from your flock, it is believed, are the only ones who may enter. You view God as an exclusionary, prejudicial beast who shows no love or compassion toward "outsiders" because they are wrong and you are right. Those who are "wrong" are punished in different ways, depending on the religion, and believed to be excluded from the fruits of God. Those who are "wrong" are viewed as sinners, whether they are from the particular religion or not.

You are not evil sinners in need of punishment. You are all God's children, in need of love. You can get this love only from each other, along with mutual respect, and when you accomplish this you achieve, above all, self-respect. None of this can be accomplished without balance. One way to bring balance into your lives is to shift the focus you have of yourselves from the concept of original sin to the concept of original goodness. This is not to say that Christianity is a bad religion because it holds a belief in original sin. All religions have flaws. The point is that little focus has been directed toward the basic goodness of people, in a foundational sense among all peoples and faiths. There is good in everyone. If you focus on nurturing the positive rather than trying to "correct" a negative, amazing things can happen.

It is believed by some that sexual union is a sin – with the inescapable result of any offspring inheriting the "disease" of sin to spread to others. The only real way this could happen is through genetics, but this is not a genetic problem. When your scientists decoded your entire genetic blueprint, no "sin gene" was discovered. Nor will it ever be. If the sexual act itself is the cause of sin, you could solve it. If you were to clone yourselves or use other forms of DNA manipulation to create infant offspring, without the involvement of a sexual act, then you would still have human beings who would sin in their lives. They would not suddenly be perfect. Therefore, through basic logic you can conclude the following: You have your flaws, or sins, because you are human; you are not human because you have "original sin."

The idea of original sin made a savior necessary. The only way to be free from your sinful natures would be through a savior. It made Christianity "exclusive," which all religions express in their own separate ways. The only difference you must come to accept is that the savior you seek is within yourselves. It is not an outer figure who will make a physical return to do the actual "saving." *You* must make the effort. You must stop sitting around and waiting for someone else to come down and solve your problems; otherwise, they will never be solved. Waiting around for over 2000 years without any results should be your first clue. Never in the history of mankind has the accepted "savior" of any religion (out of the many religions that have believed in one), returned in a verified, believable way to perform their expected "saving" duty.

All of your religions are sinful, just as humans are sinful. They are guilty of "falling short of the mark," which is what the word "sin" really means. If any of your religions provided the answers that you have been seeking, then their amazing and miraculous success, if experienced, would have spread like wildfire. Humankind would have abandoned all other faiths and converted to "the truth." But no religions have done this. They have all fallen short. They are sinful. You are no more sinful than your religion is. If you break free of your religion and seek only the truth, you may be in a better position to find it. This idea is incomprehensible to most of you at this time. You need something to believe in, but believing in nothing and starting a search for truth from scratch could be the most enlightening experience of your lives – should you know how to proceed. By doing so, you may not become free of sin, which is part of your nature, but you will most certainly become free of its most active agent, meaning religion.

The man-made idea of a savior and the promotions and promulgations thereof, keep many people from giving up and leaving the flock. When you approach the job of the savior *as coming from within yourselves*, without a specific religion dogmatically controlling the belief, your work will truly begin. Those strongly connected to their chosen religions will be aghast at this idea, which merely shows how far you must go to achieve real peace and progress in your world.

I view your religions from a place of detachment. Your spiritual progress is not dependent on your religions, although you make it so. For example, from an outside, clear-seeing observer, here is a literal definition of your religion called Christianity.

Christianity: The belief that a vicious, vindictive God who is supposed to be "all-powerful," sent his only son down to your world instead of coming himself so the son could make you live forever if you symbolically eat his flesh and telepathically assure him that you accept him as your master, so he can rid you of an "evil force" called sin that is claimed to be passed down like a virus in humanity, generation after generation, because a woman made from a man's rib was convinced to eat fruit from a magical tree by a snake who could talk, with all of subsequent humanity being punished thereafter by an "all-loving," "all-knowing" God who could not prevent it from happening.

Other religions have similar mythical stories – so cast them aside for now, and let us go straight for the truth. As you will soon discover, irrational belief structures are not as powerful as direct, inner experience.

Opposites and balance are important in your search for truth. The world contains opposites in every form: good and evil, night and day, up and down. Your mind creates the concept of polarity in them all, and forces you to see the world in a dualistic way. You might argue that the world naturally contains up and down, dark and light, so you have no choice but to perceive things this way – as opposites. This is true, but only to a certain extent. The world must clearly have enough disparity within it so that things are different; so that everything is not the same. For example darkness is simply darkness. It just "is." Light is simply light. But your minds operate with polar opposites so you project that into the world. It is very difficult for you to see and understand this fact, but it is true. Engaging in this charade has caused you to create an endless parade of "enemies," including your own higher Self (which you fear), and has distanced you from God. The answers are not "out there."

You do not create the outer world, but you perceive it in a warped fashion. You do not "create your own reality," as some of your New Age teachers have claimed. It already exists around you, it will exist after you die, and you do not create it as you go along. But you do create a *limited framework* of that reality by viewing it through a form of polarized dualism that creates your enemies and most of your other problems. If you can begin to see and act in holistic ways rather than constantly battling each other through limited dualistic thought patterns, you could then begin to *change* your reality, which is within your power, rather than *create* it, which is not. In other words, the reality and form of this world will always be there, but how you live in it is up to you.

Why can you not create your own reality? If you are connected to God and God is the greatest of creative powers, then why can you not be like God and in effect be, at this very moment, co-creators in this reality? I will tell you why and then, when you finally understand this, it will be time for you to get off your high horse and start treating nature and your fellow humans with respect.

If God were truly a creator, the argument would make sense. But God is a *manifestation*, not a creator. There is an important distinction to be made here. God is the Oneness of all things. He does not create anything extra, or anything new. All that exists is simply a manifestation of the One. Reality is energy and appears as a series of vibrations that gets used in different ways. It never becomes bigger or smaller due to the addition of "created" material. The material, all of it being God, simply manifests in different ways. There is only one Absolute, and it is everywhere. It is distributed in various forms and energy levels.

God did not create Himself. If He were a separate entity from this reality he could have created it, hypothetically. But He is *not* separate from this reality. He is everywhere and in all things. You have spent centuries harping on the great creator God, speaking on this aspect as if it were a foregone conclusion. He did not "create" you out of nothing. Nor do you create anything out of nothing.

You are manipulators of existing energies rather than creators. As you produce your various inventions – many for the sake of exploiting others for short-term profits, or at the expense of the environment – remember one thing: it all belongs to God. None of it, despite the illusion of the "power" that you cling to, belongs to you.

God is immortal. I am immortal. You are immortal. Your bodies are not. You are merely taking part in the great manifestation of God. God was not created. He has manifested Himself as this reality, so it is impossible for anything within it to be a creation. It is all simply One. It is the Whole, or it is nothing. Those of you who have awakened know this and have experienced it. Reality is based on consciousness, not on matter. Your greatest scientists, after years of pursuing matter as the foundation for reality, have begun to realize that consciousness is the foundation they seek. It never breaks down like matter. When matter breaks down it does not completely vanish – another manifestation may be born from its energy. The key to reality is this field of consciousness, the great manifestation of God. Reality was never born, it never dies; it has always been present. And so have you.

I am the immortal Overseer. I know this. Explaining God to you is a challenge, because you have displayed over the centuries a complete lack of understanding in this regard. Your conduct has been a complete and utter failure if you have been trying to display an understanding of God. You have used the idea of God to murder and manipulate your fellow humans for personal gain, with violence and personal gain representing the least desired aspects of a true understanding of God.

Those of you who believe that the Earth is nothing but a "prison planet" for souls that have been cast down and isolated for their discretions are not far off – only because your world has become a prison of your own making. Ignorance is your biggest enemy, other than being blinded to the realization of your ignorance through your own arrogance. If you could see your ignorance, you would do something about it. But often times, your arrogance will not allow you to acknowledge your errors.

An understanding of God can set you free from this trap. These words have started you on this path and the following exercise will bring it more fully into your life.

Exercise

Introduction to God

This is a simple three-part exercise. There is no other way to form a powerful bridge to God. It is much like building a bridge, which is done in steps.

Whenever you first meet someone, it should always begin with a proper introduction. A higher Self, or form of higher consciousness, is completely alien to most people. Those who experience this state for the first time consider it to be, without question, a life-changing event. So if you know that such a powerful event is harbored within you, it would be best to introduce yourself to it – if only for the purpose of acknowledging it and creating a first step toward a meeting. After all, someone whom you've never acknowledged is not likely to come knocking on your door for a visit unless you first extend some type of invitation.

Part One is meant for this purpose. In a certain sense it acts as a *rivet*, being a bolt or pin usually made out of steel, which holds the heavy beams of a bridge in place. This invitation will bind everything together that will follow, like a rivet to a bridge.

Part One: Take some time and extend a prayer to your higher Self. Just acknowledge it. Thank it for looking over you and guiding you for all these years. Because it has done exactly that. Most people have prayed to God their entire lives instead of to a higher Self and there is nothing wrong with that. There is a God and He is there. However, there is a part of you that has a more direct connection to God than you do. It is your higher Self, or the soul. If you create a direct connection to this part of you, you create a more direct connection to God. But many of you think you can somehow bypass this part of you and go straight to God.

Thinking this way is like having your best friend (higher Self) on the other side of a raging river. He is trying to shout instructions to you on how to get across but you cannot hear him. He jumps up and down and waves a big flag with your picture on it – with you crossing a bridge – trying to get your attention. For some reason, you do not even see him. This "best friend" is really part of you. But you do not see him. Some of you will notice him and will want to get across, but you do not know how to get there. So your friend grabs a rivet, heaves it across the river and hits you right off the head. Instead of knocking you out, it wakes you up. Now you know who threw it! You pick it up, identify it and start thinking about what to do with it.

This prayer/acknowledgment will help you make contact with this part of you. From there, in Part Two, you will learn how to build on the connection and ultimately, in Part Three, traverse the gap that exists between you and the hidden spark of God that quietly dwells within you.

No one can know God unless you first know yourself. That is why we have previously covered the "know thyself" material. It takes time and work to truly know thyself, which you should have invested in by this time. You should be much further along on this path than you previously were after completing the material and exercises involved. If that is the case, you are indeed ready to progress.

Let us start with a basic preliminary understanding for Part Two. What keeps you alive from minute to minute is your breath. You cannot stop breathing for very long or you will die within minutes. It is true that you also need and food and water to live, but the most essential key to your life is your ability to breathe. Your consciousness cannot function without it.

There is an important connection between your breath and your soul. The Greek word *pneuma* not only meant "breath" in the ancient world – it meant "soul" in a religious context, to the ancient Judaic and Christian religions. In the East, the Sanskrit word *prana* means "breath," and also equates to the universal life force in your body. This double meaning among cultures is no coincidence; your breath connects with your soul in the body, so the breath and soul were often considered one and the same. If you lose the ability to breath, your soul will leave the body and your life force will be gone – it's that simple.

Your breath not only connects with the soul, it is intimately connected to the mind. An agitated person, for example, breathes quickly and excitedly, while a deeply relaxed person breathes in a very shallow fashion. This example reveals one of mankind's biggest problems. It shows that you allow your emotions to rule over you. You have been allowing the state of your mind to control your breath – however, *your breath can control the state of your mind*. And the state of your mind can reach very high levels if you know how to do it. It is always better to control your emotions rather than having them control you. When you succeed in this area it will be a huge step in your evolution.

Your mind depends on the oxygen you breathe to receive enough blood for the function of thought. When you are agitated, much blood and energy flow through the brain. When you become relaxed your mind will think much less, if at all, so requires less oxygen. You can relax the mind into higher states through your breathing alone, instead of trying to force the mind to relax by trying to control your thoughts directly. Those who claim to have had life-changing encounters with God or experienced this force in some way have always been in deeply relaxed states prior to accessing this level. Your thought is a *barrier* between you and God. So a vital step toward God requires a diminishment of thought.

The best way to introduce your conscious self to your higher Self is through your breath. There is a certain energy that your soul receives through every breath you take. Some call it vital energy. Just as food empowers the body, the breath empowers your soul and thereby creates a certain energy that can be used.

Doing Part Two is like assembling the beams of a large bridge, held together by the rivets.

Part Two: Sit in a relaxed position, upright, with your spine straight. Close your eyes and take relaxed breaths through the nose. Hold them in for a brief moment and experience the air within you before exhaling. Give your full consciousness to the breath – it is keeping you *alive*, so be aware of this. Live *in* and *with* each breath you take. Become one with the function of breathe. If you become uncomfortable or are disturbed or distracted for any reason, stop. Start again only when you are relaxed and are quite certain that you will be able to resume undisturbed.

Slow your breath down in order to invoke a deeper state of relaxation. Be sure to remain comfortable as you slow your breathing down; you should never have to struggle for breath. Although slow, it must be natural.

Your thoughts should diminish as your breath diminishes. This will remove the barrier to your soul. You are not conscious of your souls because your minds thoroughly dominate you. Part Two puts the mind aside. Your minds have fooled you into thinking that the soul should be seen like everything else, rather than *experienced.* You cannot hold up and display joy or happiness any more than you can the soul. It is not like a spider or a frog that you can capture and physically examine. It is something to be experienced – but not with the mind.

Once you are substantially relaxed and not bothered by thought, move your awareness into the center of your chest, where the soul resides. It is not in your heart, nor in your lungs, where you have been focusing in relation to the breath – but is between them. The soul chooses to occupy a point in the exact center of your body, as you will one day verify in a "scientific" sense. Its' duty is to radiate its energy outward, on equal terms relative to the body so must do it from a central point. You should feel the presence of the soul by focusing here. Do not think about it, just feel it.

If you can feel it, you have now connected your "rivets" to your "beams." You have built the bridge. It is now time to point it correctly and traverse it. You may prefer to view it as a ladder, but either way, it can be your direct connection to what many will consider God. This has been the goal of religions for centuries, and now you have a simple three-part exercise to accomplish this aim.

Part Three: As you breathe inward, move your attention to the your soul-force in the middle of your chest. View it within your mind's eye as a small luminous circle of living energy. With each breathe, feel your soul essence expand within you. Visualize this luminescence expanding, ever so slightly, with each breath you take. It is in the middle of your chest for this very reason, to be a harbor for your soul and the central point for its energy – but now, for the first time, you are consciously creating this energy. It increases with the breath because the breath energizes the soul.

Your soul energy is composed of the same God-force that drives the universe. You have that force within you, so all you need to do is connect the energies together. When you feel the energy welling up within you, visualize your bridge or ladder going up through the top of your head, where your highest energy center, or chakra, is located. Take a deep breath and send your soul energy in a slow but steady stream up through this opened avenue. When you shoot your soul power up through your heavenly ladder, it will know *exactly* where to go. Your higher Self awaits it. You do not have to "direct" it anywhere except straight up and out. It will find its' home in the universe and align you properly.

Some of you may feel the connection being made; others will not. That is not so important. What results in your consciousness is what matters, and the spiritual growth that you will attain.

Do this exercise once a week to keep you on your path. Keep in mind, however, that this is not a magical formula that will immediately change your life. You still have free will and have important decisions to make every single day. Your intuition, or "still small voice" will surface more often, usually as a "feeling," however, rather than a real voice. It will be easy to ignore it, *but do not*. Act upon these feelings or hunches. This is your guidance. Eventually you will find that decisions become easier, whether they involve intuition or not, and things happen more for your benefit when others benefit as well. Everything connected to you becomes aligned and works better because you, yourself, are aligned. When others get on their proper paths with you, groups and societies will begin to transform. Humankind will be on its way to a better future, instead of wandering around aimlessly in the dark.

This entire exercise should have created a bridge between you and God, with the higher Self as intermediary. If this does not happen right away, your patience and practice will make it so. You will be fully "aligned" as a result. When your automobile is out of alignment it becomes dangerous and does not steer straight. You have all been out of alignment, steering off course from nature and crashing into each other in violent ways for centuries. With the proper practice and awareness, this degree of human violence may no longer be necessary.

Understanding Your Relationship to God

The previous lesson should have helped to create an initial "bridge" between you and God, through the use of the higher Self. This lesson may strengthen this connection. It will give you a better view of mankind's relationship to the divine power in the universe and provide you with a more personal connection to it through the exercise at the end.

In order to understand your relationship to God, it is important to think clearly from the very beginning of your search. Before you begin exploring your thoughts, questions and ideas of God, I will define your terms. A basic law of philosophical thinking is to define your terms first so that everyone is clear on the meanings being conveyed and no confusion results.

Humankind would like to *understand* its relationship to God. You have been trying to do this for centuries, with little success. The word "understand" is divided into two parts – under, meaning below, and to stand. So to fully grasp the meaning of something you must virtually "stand under" it – so that you see its basic *foundation* and what it is truly built upon.

The word "relationship," as in Understanding Your Relationship to God, is structured the same way. You have the word "relation," which means a close connection, and the word "ship," which means to carry forth. So a relationship is to carry forth a close connection.

The word "your" in this context – Understanding Your Relationship to God – refers to humanity, rather than you, personally. But a personal understanding is your goal as the reader, and must be accomplished first, before the whole can do it. I present the collective view in this lesson so you can see what the outcome and goals of humanity should be, while the exercise at the end of the lesson is meant for you, personally.

"God" is the most difficult of terms to define, because everyone has their own conceptions and experiences of the divine – varying from pure atheists who find little to accept or work with, to radical fundamentalists who devote their lives, sometimes literally, to the imagined and extreme "causes" of their deity, to spiritual masters who have touched the very essence of deity itself. This disparity among you is why the word "your" is being used within the context of humanity, as a whole, rather than personally. This is the only way to address it in general terms.

You are all part of God. You all share a connection to Him. Although God is not a male personage, I am using the words "He" and "Him" as the figure of speech that you have been most comfortable with. God, however, is an all-encompassing force and consciousness that connects all life forms together. That is the simplest way to present Him to you in the context of humanity as a whole.

With our terms defined, we can move into this important lesson.

With all life forms connected together through God, it should be no surprise for you to learn that nature plays an important role in expressing God's work on the Earth. You, as humanity, once played a bigger role in nature. You were once more in tune with it so, as a result, you were more in tune with God.

When you broke away from nature, you broke away from God – not completely, but you have made it more difficult to maintain your connection. The Garden of Eden story is an allegorical tale of this original separation. Today you consider yourselves "above" nature rather than part of it. Your giant egos have sat you on high, in the virtual seat of God, displacing Him as you rule over nature and attempt to control it on all levels. You are, however, operating as "false Gods." It is God's job to control nature, not yours.

Among your religions you are always on the lookout for false Gods, fearful that they will appear. The next time you stand before a mirror, you must take a closer look. You should actually *speak* into the mirror and assure yourselves that the false Gods are here. They operate as your own worst enemy. It is a tough pill to swallow, but you will never make any progress until you acknowledge the fact that the false gods are you.

Prayer is common in many of your religions. Most of the time, however, it seems your prayers are not heard or go unanswered. The reason you pray to begin with is because you have been separated from God and are trying to restore the connection to Him that you have lost. To get what you ask of God, you need to know that He hears you. You need that connection first. So you pray. In so doing, you are focused on something you believe to be outside of yourself. But your higher Self, found only within you, is the key that you seek. Your higher Self holds your connection to God, but you are so focused on some kind of fictitious man in the sky, that you are missing the entire boat. It is your higher Self, alone, that can get you on board this boat. Establishing a relationship with your higher Self will not instantly solve all of your problems, but if you can succeed in the modest goal of initial contact, you will finally be moving in the right direction. It is not easy to develop a powerful, working relationship with the higher Self. It will take time and the same form of devotional practice that any religious activity would normally entail. If you are not willing to devote yourself to the time and the practice, you may not achieve the spiritual power and connection to God that you seek. Each of you has a higher Self and all of them, collectively, bind you to God quite closely, although you are not conscious of it.

Each type of life has its own collective unconscious. You can study nature and see it play out. All of the birds know, without logic, how and when to migrate. So do the whales and other forms of sea life. They don't need to think because they are tied in to God. Bees work as a unit within their hives – without logic knowing exactly what each needs to do, so that collectively, as a whole, they can survive. Ants, termites and millions of other insect types simply *know* how to live their lives.

You no longer know. You have thrown it away in favor of "progress." Your collective unconscious is still there, like a shadow, but you deny its existence. Do you not think that all other life has this important feature, but that you do not? Do you really believe that you are so much "better" than the other life forms you share the planet with that you have the right to literally destroy them, with your blind actions? What is it that you fear? Maintaining your intuitive instincts will not relegate you to a primitive animal existence. Refining these skills are important. But you have run from your nature in a desperate attempt to leap beyond your humanity to play the role of gods, catapulting yourselves into a void by using a warped form of "logic." You fail to see the darkness in the void where you have landed.

Instinct, however, is not blind. It has great wisdom and is a protective guidance mechanism. Your lower forms of instinct are still intact, like caring for your young and the urge to procreate. But you have a higher form of instinct that should be guiding your survival. You have turned it off. You have failed to develop it so you are spiritually stunted. Your egos have grown so large that your individual pursuits have become more important to you than your collective survival.

You are like the prodigal son who has left home and must learn some hard lessons. Only after realizing the error of your ways will you make a conscious decision to return. But by then, it may be too late. You have abandoned your own family – meaning the other Carbon-based life forms you share the planet with. They are nothing more than commodities to you, to be bought, sold or butchered until they are brought to extinction, or to its very brink. You have denied your own mother – the living Earth herself, Gaia. She is alive and depends upon you to live in balance with the rest of her intricate parts, but you have turned your backs on her and act like a cancer, eating away at the delicate fabric that has been otherwise woven with such care.

It is true that you are beginning to awaken. You are beginning to think and act holistically. This is a sign that you are starting to turn back toward "home" by getting in tune with your collective unconscious. It can be observed in lesser but more balanced life forms, but when a higher life form abandons it, the entire ecosystem suffers *despite* how well the rest of it may otherwise work. Your foolish deeds reverberate throughout all of nature.

For example, your bees, that know how to live so harmoniously among themselves and within nature, have been mysteriously dying off worldwide, to the point of creating a dangerous crisis. They are needed to pollinate virtually all of your flowering plants, including essential food sources that you and many other species depend upon. Your bees have been exposed to an explosion of toxic pesticides that kill them and genetically modified plants that were created in laboratories instead of in nature. You have poisoned the bees, and you are poisoning yourselves. To be in tune with God, you need to be in tune with the other life forms that you are supposed to be not only sharing the planet with, but protecting. You cannot step in and take the place of nature by creating synthetic foods of your own, just so you can patent them, own them, sell them and make money. Many of your greed-driven schemes are insane. Your plans to put nature out of business will severely decimate you within four generations if you continue on this path. You have enough knowledge to reverse this trend now, but not enough wisdom. Cultivate it. Fast.

Making these mistakes serve one good purpose. They open your eyes to the folly that you create. The only way you can listen and learn is to act with such blatant stupidity that it rises up and slaps you in the face so hard that you finally, *finally* have no choice but to listen. There are some very big "slaps" waiting on the horizon unless you can heed my words and adjust your actions. That is the purpose of this book.

God is there, but He will not make the right decisions for you. He has given you free will. For those who know that God is there, intuitively or otherwise, atheism is not an option. Nor should you be acting like gods yourselves, because you have not cultivated the proper wisdom. There are consequences for your actions. God is truly there in the world, but in a non-traditional way that must still be officially "verified" by you to be believed.

Many powerful people act with impunity. They believe that without the direct presence of God being evident in the world, it gives them free license to act without morals and to plunder from the masses. Power and money have put them in an advantageous role that requires answering to no one – except for the other "players" in their high level game of planetary power.

Acting like gods puts you further away from the true Source. You are not ready to meet God because you must meet *yourselves* first. But your relationship with God exists. It is there, within each and every one of you. It will be strengthened with every moral act that you perform. Kindness and compassion are strengths, although you have been treating them as weaknesses for centuries – jumping out of the shadows like predators to exploit anyone who might exhibit them. Violence, brutality and cunning deception have ruled the day because such actions bring spoils and riches to your door. These spoils have blackened your souls. You have lived in darkness and have failed to see it.

The time has come to shed a stronger light on your lives and to see them for what they are. You have started to awaken, and must be encouraged to continue to follow this light. This light – your true guiding light – comes from God and is within each of you. Only a few enlightened souls have experienced this reality to date, and mainstream society has been quick to ridicule them because they are so different from the rest of you. Their teachings get ignored. The enlightened confuse you; you do not know what to do with them because they do not fit in. Yet these are the ones who have found God.

As you continue to evolve, more and more of you will become as they have become. But for now, you must be content with working towards that end to the best of your abilities.

Exercise

Awareness of God

Your consciousness is like a beam of light. It shines out into the world so that you can understand your physical surroundings and work within them. This beam of consciousness does not shine upon its own source. God is the source. You merely use the "light" as it projects through you. If you can *expand* your consciousness beyond this thin beam, you may reach back into the realm of God and understand Him better. God is the source of your consciousness. He is also the source of your world. Everything is connected through a giant harmonic web that vibrates at different levels, including yours. Therefore, God is looking out upon the world through your eyes, and through your consciousness, continuously connecting to all that you encounter. Everything is connected and it is through *consciousness* that everything is connected and evolves.

You have taught yourselves over the centuries that God is looking down upon you; that he is watching you and judging you. It is time for you to move beyond this outdated view and into the next level of consciousness. You are ready to reach the clear understanding that God is part of you, and you are part of God. Therefore, God does not look at you or your world in a completely detached manner. He views the world *through* you.

Your exercise is to gaze out upon the world not as you normally perceive it, but with the knowledge that your consciousness is the consciousness of God. It is the simplest of all the exercises in the course, but could well be the most powerful. Study everything on behalf of God, who relies upon you for Him to experience your world. God is not a physical being. He uses the collective consciousness of mankind to experience your world in countless ways. He does not control you, however. You are an emissary who has been given free will. Therefore, all of your choices are important. You have an obligation to live in Godly ways. If you fail for any reason, you are still in the world to learn from your mistakes and evolve. You are moving back toward God as you traverse through your world and confront your challenges. As you progress through life, you should be moving toward a moral Godly consciousness.

If you do this exercise daily you may soon feel the presence of God within you. You may even feel this power the first time you try the exercise. It all depends on how far you have evolved and what kind of connection to God you have been able to construct up to this point in your life. One thing is certain. Despite its simplicity, this exercise is capable of building a stronger connection to God if you maintain its practice with diligence.

Many of you will try this exercise a few times and then move on with your lives. You will experience the novelty of it and then fall back into your old ways. *Do not underestimate this exercise.* While you do it, there is far more going on than meets the eye.

This lesson will have created a stronger bond between you and God, just by reading it alone, for the very first time, without even doing the exercise. There is true power in that. The exercise, however, will strengthen your bond with God each time you do it. Many benefits will result; some that you could never imagine. If done properly, one of the benefits of this lesson is that your path in life will be clarified and become more secure. Work with God and God will work with you.

Lesson Eleven

Enlightenment

This lesson does not guarantee you enlightenment. Each one of you is on a unique and different path, individually. You are functioning at different levels and advancing when ready, despite the fact that your final destination is the same for all. Some of you are much further along on the journey; others have more to learn and more obstacles to overcome.

This lesson will clarify your personal vision regarding your journey through life. You may see "the bigger picture" more clearly, which will allow you to *advance* toward an enlightened state should you be so inclined.

If you can understand your relationship to God in a conceptual sense, then you have done well with the previous lesson. You will have a much better view of your place in the world. But a conceptual understanding pales in comparison to having the direct *experience* of God. You do not have to learn a great deal about God and gain a conceptual grasp of the divine before having such an experience. An experience of God can come out of nowhere and can happen to any one of you.

However, few normal, everyday people have such an experience. It has been known to happen but it is far more likely that such an experience will come to those who work at it. They understand that God is found within and that He does not sit up high in the sky, looking down on you to judge you or punish you. The true God does not spend His time judging or punishing. His energy is directed within a form of *guidance*. Humankind as a group has not learned yet how to listen to this guidance because it comes on a very deep, inner level.

Spiritual work on an inner level can bring you closer to God, so you can not only experience Him, but access His valuable guidance. This has been known by your great sages and mystics for centuries. They do not run through the streets proclaiming this great news because they are far wiser than that. They know that each of you have your own paths to traverse and only when you are ready will you cease to look outward for your satisfactions, and begin an inner quest.

Starting this inner quest in earnest shows that a major *shift* has occurred. Your focus, and therefore your ideals, has moved from the visions of materialism and the incessant desires of the ego to something more deep and meaningful. You will begin to express this deep meaning and your awareness of it in the outer world while you progress on your inner journey. This is the "shift" playing itself out.

The end result for some is to experience a state of enlightenment. This, however, should *not* be a goal. Many of you chase enlightenment like it is an elusive creature that, once captured, will reveal its secrets and give you everything you need in life. No one, absolutely no one, who has ever chased enlightenment with this mindset, has ever "achieved" it. One of its secrets is that it is not something to be achieved. You already have it. There is merely a veil or web that covers it over. You must learn how to throw back the veil and reveal what you already have ownership of. It must be claimed but, like a great treasure, you need a "map" to find it. This lesson may help you, and may even serve as this map for some of you.

In the meantime, enlightenment waits for you like a great wise avatar, a higher Self that watches and observes. It waits. Do you want to eventually die and leave Him or Her still waiting? Or do you want to find your most valuable asset while you still have time? This state of being and your relation to it has a lot to do with why you are here, but most of you do not know this. It is time for you to know this, and to take some further steps on your journey.

Kundalini is a hidden spiritual power that rests at the base of your spine. Years of extended spiritual practice often consisting of meditation, yoga, or a combination of both, has been known to create a spontaneous rising of this energy, up the spine, and through the various energy centers in the body – sometimes referred to as "chakras." Upon reaching the crown chakra, locating between the eyes, it will "explode" in your mind and illuminate it. This is one reason why they call it "enlightenment." An immediate shift in consciousness comes with the explosion – you will understand the Oneness in the Universe because you will become dissolved into it and *experience* it firsthand.

This is a life-changing event for those who end up experiencing it while still in the body. These people are often transformed, and old, material-based ways are cast out. They have been known to become artists, writers, teachers, composers, humanitarians or key figures in holistic based movements, to name a few areas, after previously having little to do with these pursuits. Many convert to vegetarianism as well. In virtually every case, the old life being lived up to this point cannot possibly continue. Your great spiritual masters have, in almost every case, experienced this state of mind. With the chakras opened up, a free flow of powerful spiritual energy courses through their bodies. The normal blockages of this energy, found in most everyone, are no longer there. Some of these spiritual teachers or gurus have been known to pass this energy on to chosen followers, when they have determined that such a follower is ready to experience a glimpse of this power.

Those of you in the West who do not observe Eastern religions have been conditioned throughout the centuries to believe that such experiences are evil. Accessing your true nature and experiencing it is not evil. Experiencing the God-like part of you is a major feat, accessed by those who are ready for it. Those who are not ready for it are fearful of it, and call it "evil" rather than confront it. You had a great psychologist who understood the human mind as well as anyone whose name was Abraham Maslow. He once said to the rest of you, "We fear to know the fearsome and unsavory aspects of ourselves, but we fear even more to know the godlike in ourselves."

You will create positive change in direct proportion to the progress you make in conquering your inner fears. But many of your religions have conditioned you against exploring these realms. They provide a safe harbor, shielding you from uncharted territory that is begging to be explored. You are happy, docked in your comfort zones and do not want them to be disturbed in any way. You do not care about evolving because religion lets you cling desperately to your beliefs in the face of new discoveries – discoveries made specifically about yourselves, your true natures and your hidden potential. Your mantra has always been, "Don't bother me with the facts, I have my beliefs."

It is time for you to study consciousness and what it really is. What you learn may start you on a new journey that will change your life. You must first be willing to hoist your anchor and sail out of your comfort zone. It will be an inner journey rather than an outer one.

The outer journey has led you in circles. It has always been a financially motivated, politically based façade, tied in with strong religious principles to cement it within your cultures. You have a deep-seated need to believe in something spiritual, so religion has been there for you in a basic sense. This need is part of your psychology and those who know this have exploited you for their own agendas.

You have grown, however, and will soon graduate to something deeper and more meaningful. You now have a choice. You can continue to be exploited or you can go on an inner journey. You can embark on something deeper that will bring you closer to your true Self and thereby closer to God. I am not suggesting that you abandon your religions. They serve a good purpose and are a starting point for virtually everyone. For example, Jesus told you that the kingdom of God is within you, but the authorities of his religion forbid meditation, which is a clear inward avenue, and will tell you it is evil. Listen to Jesus, the founder of the religion, and forget in large part what came later from those who added to it. They encourage prayer, but prayer often relates to the outer world or your selfish desires and does not *quiet the mind*. The mind continues to ramble when you pray. These rambling thoughts form a *barrier* to God rather than a connection, which is the exact opposite of what you seek. It should be no wonder to you that so many of your prayers go unanswered. To connect to God, a deeper search is involved.

No sanctioned form of Western meditation has been developed around Jesus because the church must be the authority, rather than you being able to find it within yourself. But Jesus clearly wanted you to look within, otherwise he would have never told you about it. So do it. If you are aware of the exploitation that has occurred over the years to your faith and, as a result, to you, you can then operate within your faith with your eyes completely open – but looking inward. You will gain more understanding from your chosen faith – not less. You will know, instinctively and more accurately, which teachings ring true and will have value on your new, inner path. An awakening is at hand. You are a powerful spiritual being. You do not, under any circumstances, have to be nothing more than a "follower" to express yourself spiritually. If you remain only as a follower, you will not grow as much spiritually. And growing in a spiritual way is a big part of why you are here in this world. You did not come here to be a consumer. You did not come here to be a pawn on the global economic chessboard, to buy what you are told to buy and to do as you are told. But the vast majority of you people are doing this exactly. You are sleepwalkers, completely blind to your spiritual natures.

It is time for *you* – the individual reading this book – to awaken. And I mean truly awaken. Something within you is stirring. It is *responding* to these words. You can finally hear it, in an intuitive way, and finally, *finally* you can understand what it is telling you. It is a small voice and it does not use words. But you can understand it enough to know that it is time for you to awaken. After getting this far in reading this book, you have now found out why you have bought it. This book, however, is *not* your key. Your key is found only within you and you have contacted it, just briefly. This book is only a tool. It points to the key. Now use it.

Strictly religious people do not exist within an enlightened state. Becoming enlightened is a spiritual experience rather than a religious one. The difference is that a spiritual experience comes from within you. It is more powerful than a religious one, which results from following man-made guidelines. You will not find anywhere in your world a fully enlightened fundamentalist extremist. You will find, on occasion, *former* fundamentalist extremists who have become enlightened. You cannot be both, because one transcends the other.

Those who have been awakened have had their minds opened up. They exhibit more compassion and understanding. They realize that the chosen form of religion they have lived with and observed during their entire lives is not the only answer. For example, it is often said by those who have experienced God firsthand, that it gave them the most incredible feeling of *peace* that they have ever known. This peace is so extreme that its description is beyond words; it must be experienced to truly understand it. Most religions, on the other hand, engage themselves in *violence* for the sake of God. This has always been the case in your world. The sad fact is that more people have been killed, maimed and tortured on behalf of religion than for any other reason in the history of the world. Your religions are filled with violent acts and teachings, found in holy books or passed down through your rituals. People claim that God has sanctioned their violence because it justifies their otherwise unbearable deeds. However, God was not a true part of your wars or their supporting religions because you do not have *peace* during wars and you do not teach it properly in your religions.

If you kill someone for "God's glory" in war it is considered a blessing, but committed at any other time you treat it as cold-blooded murder. Today there remains little peace in your world. Violence begets violence; you cannot achieve peace through violent means. You must be peaceful from the start. Your world is still so barbaric that you must respond with violence or be yourselves destroyed, so you must operate through self-preservation. Eventually you must come to a spiritual realization together. You will fight amongst yourselves like children until you tire so much from your carnage that the realization will begin to dawn on you – at least as a majority. This is where you are now. There are vicious pockets of ignorance still left in your world, however, and you have a great challenge ahead. The more ignorance you have, the more dogmatic the people. The more dogmatic the people, the more violence and extreme actions there will be.

The more peace you have in the world, the more you can say that you have truly found God. Your progress with God can be measured by the level of peace you have in the world. But first and foremost, your job is to cultivate peace within *yourselves*. When you find this inner peace you will have found what religion strives for. There are many ways to discover this. Once done, you may be able to move beyond your religions, which have *divided* you. You may become more spiritual, rather than religious. Religions create followers, while spirituality reveals the God-force within you. You will no longer be reaching out to God, but reaching in – and thereby become more successful in attaining your inner and outer peace.

In a general sense, Eastern religions are more spiritual in nature. A strict Western approach to your inner advancement is limiting. Despite certain religious teachings from various faiths worldwide, it is not a blasphemous crime or a sin to broaden your horizons. It is just as "sinful" to prevent your own growth by remaining strictly subservient to your faith. There is really only one true religion and it is found within you. Your great founders of various faiths chose different forms of expression, which is a good starting point, but it is now time for you to express *yourself.* This is your life, your search and your God – to find and love and to *experience* for yourself directly.

Religions, in relation to your spiritual growth, have helped you to stand up much like a toddler does for the first time. But you must continue to grow. Your spiritual development needs to keep pace with your intellectual growth. A child cannot simply get up and walk without understanding where it might go. Without such knowledge, a child can get itself into a great deal of trouble. That is where you are now, in a global and societal sense. You need a better spiritual understanding, and *wisdom*, to get you out of this troubled area that you have wandered into. The way out is to travel inward, so you may retrieve what you need and continue your journey with wisdom, compassion and renewed understanding.

Meditation and yoga provide two powerful means from the East toward enhancing your inner awakening. Some people who have experienced a spontaneous awakening have suddenly found themselves engaged in yogic postures when the body has never previously done them. There is a mind-body connection that you have and the body "knows" how to express it when the right time should happen to arise. If the right time does not happen spontaneously, as is the case with most of you, then it is up to *you* to get things started. You cannot depend on your awakening to happen by itself.

An approach with yoga is more extensive than doing just one exercise that could follow, and would requires another book (of complete instruction) or a class for you to take to be effective – so this will be up to you if you feel led to do so.

Exercise

Meditation

As far as enlightenment is concerned, the best tried and true method used in reaching this state has always been meditation. If you do not have a meditative practice, you may consider employing one. It could take many years of devoted practice to reach an enlightened state, but there has always been no better way. The greatest spiritual masters who ever lived have in common the use of devoted meditation to achieve a deeper understanding of their relationship to God and of themselves.

In the West, meditation is not normally part of the religious approach to things and is too "foreign" for many to embrace when first introduced to the idea. Some western religious authorities warn against meditation as being "evil" or of the devil because it is not found in western religious systems. Turning inward in this way, however, is not evil because it is *you* who is being explored. God did not create humankind as inherently evil, nor are you children of the devil. You are children of God. You are allowed to be spiritual and to seek truth within yourselves. Do not mistake authority for the truth. *Your* truth is the real authority and you must be allowed to find it. No religious "authority," whether a person or a religion, has the right to stifle your search for truth.

Meditation, once again, is a good way to do this, having been proven through the ages. It is a way of quieting the mind. If you are not accustomed to quieting the mind and have never meditated, there is a more western way to approach it, which you may find easier. It is an "introspective" avenue and is good for everyone to try, whether you are a beginner or not.

Pick a positive word, any one at all, that you believe has *meaning* to you. Avoid the negative ones. Good ones to start with are life, love, happiness or success. Get into a relaxed and quiet atmosphere and think of this idea in every possible way, until you run out of thoughts. You are not to make a list of other related ideas, but to bring forth scenarios in your mind of this particular one. Your mind will begin to wander at times into completely different areas that *seem to have* nothing to do with your original thought. But it does. Ask yourself what the connection is. Explore every angle of this important concept. You are to avoid starting with negative words or ideas because they will pop up by themselves during your exploration. What is the connection? You must go deep within yourself to figure it out – or at least try to. Sometimes, later, once you are done, the connection will come to you in a flash. If no answer comes and you must wait for it to possibly appear later, go back to the original thought and start again. The point to this exercise is to be introspective on this idea and how it relates to you. Explore this word in every conceivable way until you finally run out of connections. Your mind will finally be quiet. It may take ten minutes, thirty minutes or an hour to get there. When you do, experience what is left. You should make sure that the word or phrase is still there, subdued, but no more connections exist. You have run out. Your mind will rest and become quiet by itself.

Another rendition is the freestyle approach. Start with any random thought; it really doesn't matter. Just let the mind go off on any and all thoughts it wants to explore. Don't question them or examine them. Just acknowledge each one as it enters your mind, gently put it aside and move on to the next. One after the other after the other. The mind will finally get it… that it's a non-stop circular process that leads nowhere, except to the next thought or desire, and will finally become quiet. When it does, enter into it the quiet and experience it fully.

Many of those who have meditated in traditional ways have never explored this "reversed" approach, so it is also good for them. When people start a more traditional meditative practice for the first time they are often "antsy," cannot sit still and find it extremely boring. In trying to meditate they are suddenly not doing anything that connects them to the outside world and if the outer world is all they know, they will fight tooth and nail to avoid the inner one. When these reactive barriers are experienced it shows that they have not yet found the essence of meditation. There is not a boring emptiness within you, as they will claim. There is an *entire universe* within you. You just have to put your dependence on tactile sensations aside, and delve into the hidden, inner frontiers.

The main thing is to not randomly attempt to adopt as your own the first form of meditation that comes along. Do some research first. Then try different ones. Explore. This is serious but beautiful work at the same time – it is an exploration of the true essence of *you*. There are numerous schools of meditation, many dating back to ancient times. You must find something that you will be comfortable with. Some require your eyes to be closed, with focus on the breath. Others require eyes open with focus on the breath, or an object. Some require a focus on all of your thoughts, examining them and letting them pass on without concern. Others use mandalas – beautiful symbolic artwork with a central point of focus. You should sample a few different methods until one resonates strongly with you. You will know which way to go when you find it.

Whichever form of meditation you decide upon will not manifest right away as a magical answer. You must use it with discipline, and many who lack this attribute will fall by the wayside and live their lives less richly. All forms of meditation take a good level of practice, often many years, so it is your comfort that is the focus for the moment, not your enlightenment. Your enlightenment will come later, once your inner Self determines that you are ready.

You will still reap many short-term benefits from daily practice – or even from practice three times per week. You may be calmer. You may lower your blood pressure and quieting the mind will make room for wisdom to enter more often. You may make far better decisions and be less impulsive. Your life energies may flow throughout your body more freely, providing better health and peace of mind. You may have more energy. You may be more alert. You may have sudden insights that will drastically improve your life. As these things continue to manifest, slowly over time, your higher Self will be watching. It may choose to interact with you more often as you align with it, which will create your proper path in life more clearly. Your *entire being* will be making a certain degree of progress as you go, and when the time is right for a major shift, you will know it. Without question.

This entire lesson should have moved you upward on your path, either in a slight way or a more powerful one, but in a way that has been noticed by you. You should be conscious of this shift as you start to evolve in new ways. Then, as you *continue* along with your new path, an expanded awareness should unfold. If not, be patient and commit to a comfortable practice that only you can choose, as previously presented. Things will happen, and continue to happen, when the time is right.

Lesson Twelve

Changing the World

This lesson requires all that you have learned so far to approach it. It will give you a much clearer view of your place in this world, and allow you to come away from this lesson with the ability to act and make a positive difference. The ability for you to act will be a reality, but the *desire* to do so can only come from you.

Changing the world means changing yourself first. The barbaric masters among you that have so far changed the world to their liking operate in ways that force the rest of the world to comply with their preconceived notions about how it should serve them and them alone. This is based on personal ambitions and economic/political power plays. Your entire eco-system has been almost completely shattered as a result of this self-centered folly. You are teetering on the brink of disaster and only massive changes coupled with a new mindset will save you.

The best way to change the world is for mankind to step back into a *desireless* state of mind. This would mean, for many of you, a world full of people sitting around and doing nothing. But the "desires" I refer to and wish to banish are selfish desires – those that are ego-centered and relate only to you at the expense of others. Unselfish desires should be maintained. For example, your primary unselfish desire should be to transform. You cannot transform until you clearly know what needs transforming. You must face yourselves. This course and this lesson will help you do it. You must confront the greed, selfishness and other shortcomings that fuel your cultures and personalities, rather than sweeping them under the rug or rationalizing them away.

Unselfish desires will function almost seamlessly in the coming paradigm, if you can arrive there. This may not happen any time soon, but that is what you need. The world will never change until you do. You wait for it to happen by itself without making it happen for yourselves.

Remove the consumerism from your minds and focus instead on planetary and personal health. Any alien culture coming here would immediately see how sick and unbalanced your world is. Each of you must step back and view the state of the world in this way – from a completely detached viewpoint – as though you were from another world and just landed here for the purpose of observing. That is the way I view your world. The level of ignorance you exhibit is stunning because you *know* the problems you face but still do little or nothing to solve them. You are like the lemmings that run over the cliffs in mass suicidal numbers, simply because you are trained followers, afraid of expressing an individual vision. You depend upon those who tell you that everything is all right and you blindly follow them and believe them. After all, they are the "authorities." They *must* know what they are doing.

They do not. Your great authorities have suddenly found themselves at the edge of the cliff, panicked by what they see and are now being *pushed* by you from behind, because they told you to follow them. They will never admit they lied to you. But now you are *all* on the brink of disaster, with no available wisdom to guide you.

I, looking from above, have seen it all transpire. So here is this book. Use it, if you will. Or don't. You have only a few more steps to take in the wrong direction, without listening. Then you will not need help from me or anyone else. It will be too late.

I have always had faith in your lot. I have always sided with you and given you the benefit of the doubt when decisions were being made about "what to do with you." At some point, however, you must be given the chance to decide what to do about yourselves. I cannot keep saving you. If enough of you can be awakened and start to exercise some wisdom, you can then develop a sustainable structure for yourselves and the planet. You are making some efforts but they are not yet allowed to disrupt your rape of the planet, which supports your economic concerns. A sustainable answer is so far from being a reality that my concern is beginning to be more for the planet than for you.

The only hope you have is for you to shift completely into a new paradigm. The system you have has neared the end of its sustainability and has no long-term potential left. You cannot change the paradigm, however, until you change the entire mode of thinking that depends on it, and created it in the first place.

If the potential exists for individuals to become enlightened, you must come to the understanding that societies can do so as well. Although you are far from achieving it, this is what you are striving for – and learning – in this world. You have not yet learned how to become enlightened individuals, so transforming your societies in the same way will be a much larger challenge.

Your awareness, however, has shifted. You are *aware* of what you must do, but have not formulated a cohesive plan to facilitate the shift. Your economic structure cannot change overnight, but must do so as quickly as possible. The entire shift will take time to achieve and to become stable. You must have patience and plan for the future, while doing everything you can to support all forms of holistic based actions.

You must stop educating your children to be aggressive business people who will stop at nothing and succeed at any cost. All actions have consequences. If you continue to neglect the environment or those less fortunate than you in order to facilitate personal gain, then conditions will continue to spiral downward. If you educate future generations to be holistic based entrepreneurs, then everyone will win rather than those companies or individuals who create the idea alone.

I will not map out a specific plan because you are capable of doing this and it is your job to do it. My job has been to step in only at severe crisis points in your history when all is about to be lost, in order to point you in a worthwhile direction that could avert disaster or, in the worst case scenario, to save a few of you and salvage whatever is possible when the inevitable happens.

You have free will. You cannot change the world and its structure until you change the way you think. The way you think is mired in the muck of materialism, and you must rise above this with the realization of more spiritual principles. Money is not your God, despite your best efforts to treat it that way. Those who control your economies and ultimately, all of you, are those who control your international banking and financial institutions. They have squeezed every last drop of wealth out of individual nations and their people to the point where they and the people in them have become modern serfs. The system is buckling at the seams because no country is producing enough wealth to survive on its own. You have been forced to strip your land of its valuable resources to pay your debts to the point where almost nothing is left. Most of Earth's people are tied to the land so, as a result, few people are debt free – such a thing is virtually unheard of throughout the world in your middle classes. The lower classes, without any assets to speak of, are nothing more than indentured servants throughout the world who will spend their lives in virtual slavery.

Clean water, the key to life itself, is running out amazingly fast and its' shortage is a major crisis throughout your world. Your awakening has been manifesting in direct relation to the level of discomfort you have created for yourselves. Your awakening will continue to surface as your problems continue to grow. The question remains – by the time you awaken as a species and are aware enough and wise enough to repair the damage that your ignorance has caused, will there be enough time left to do so? It is unfortunate that it must happen this way, but human nature's negative side has always spurred your growth to a certain degree. At some point, when you have gained enough wisdom, it will no longer have to be this way.

In all my wisdom, I do not know if you are capable of surviving as a species. There are many brutish and violent species sharing the planet with you at the current time, but none have come close to causing the damage that you have done – both to others and to yourselves. You may self-destruct rather soon. So many life forms on Earth have come and gone in the past, so losing you from the Earth would not be much of a problem for nature. In fact, if it happens she might even celebrate. If you perish, nature would learn from her mistakes as she has always done. She would simply replace you with another life form, a possible upgrade, integrating modifications that would, in all likelihood, allow them to fare better and longer than you have.

The dinosaurs flourished on the Earth for 150 million years. You have only been on the planet for a few hundred thousand years, and are in jeopardy of a far earlier extinction. What was it that made them so successful? The answer is that they were *stupid*. Possessing great intelligence is not a good thing because you do not have the proper wisdom to govern it. You spew out endless technological advancements, but the resulting wastes and by-products poison the Earth and yourselves. If the amount of chemicals continues to enter your food and water at the pace you now experience, your entire population will be dangerously contaminated within a few generations. You are just beginning to understand this.

Therefore there is hope. I have faith in your ingenuity. I have faith in your resolve; in your ability to meet challenges; in your ability to care for one another. You are not complete failures. You have tremendous potential – more than you could possibly know. The human soul is not a monster; it is a magnificent, magical, powerful force that you have barely begun to tap into. The human psyche and the ego can be monsters, however, and they continue to step in the way of your true potential.

You must learn how to tap into the soul and bring it into your consciousness, while lessening the influence of the ego upon your behavior. The soul is your guiding light, but you often block it. You refuse it and extinguish it. Your potential is far from being met, yet you are so very, very close to meeting it. You may one day conduct yourselves with great wisdom but for now you remain clever opportunists concerned more with individual gain. You fail to comprehend the way in which all individuals will win if holistic aims are first met.

I care so very much for your race, but must allow you to grow at your own pace. It pains me to the very depths of my being to watch you while you live in darkness. I understand the soul and how it works, but to you it remains, for the most part, an uninvited guest. I feel the pain and suffering that you have had to endure and will continue to endure, and it is almost unbearable for me. I want to step in and make it right, to ease your pain and make your lives better – but they would never remain better unless you reach this plateau yourselves. Otherwise, you would never know how to maintain it. I can give limited guidance, as I now attempt with this book, but I cannot do the work for you.

It is part of an awakening process. You are awakening in direct proportion to the level of discomfort that you experience. Your discomfort tells you that something is wrong and you search for deeper answers as a result. It is a shame that it must happen this way, but that is the purpose of suffering. You always look to the sky and pray for answers when you suffer or witness the great suffering of your loved ones, begging God for a way to stop it. But it has a purpose. Its purpose is to breed compassionate action as a way of life. Very few of you extend compassion to others because you are too wrapped up in yourselves, in your egos, and in your own desires to extend this compassion as a primary form of expression. Your suffering will not end completely, but can be lessened to an amazing degree as you continue to awaken. God will not end your suffering. He is waiting for *you* to do it. You are here for a number of good reasons, and that is one of them.

I am your Overseer. I am not your babysitter, despite the fact that many of you are indeed spiritual babies. You are here to develop on your own and learn valuable lessons. I cannot walk your path for you. Those who get in the way of your spiritual growth are often opportunists who exploit others for personal gain. They prevent you from growing so they can become comfortable and quite wealthy at the expense of you and everyone else.

The media are your biggest stumbling block. In large part they fail to report the realities of the human condition throughout the world in order to entertain you and target you as consumers. You could be giving to others and truly helping the world, but your time, money and resources are instead directed toward and consumed by an economic machine that has no concern for the poor and starving masses. Money needs to be made. If too much help were extended to the suffering masses, it would divert the economic resources of the world away from the powerful, and bring strength and independence back to the masses. Your masses must no longer wait for permission to speak, but must begin speaking. They are ignorant of their true power – all of you are.

Some of your best scientists are paid handsomely, secreted away in hidden enclaves devising ways to "humanely" weed out less desirable populations – meaning those that cannot be groomed as consumers so easily – whether it be for poor economic ability or being in geographic areas that are not as economically viable. The economic machine must remain healthy, at the expense of certain unfortunate groups. It is criminal to engage in mass murder, so a certain level of stealth and sophistication is required. From time to time news leaks out about certain viruses or diseases being created in a laboratory somewhere, often with credible evidence backing these claims. It is not easy for you to manage the exponential growth of your masses, and it will not get any easier for you as the situation progresses.

More than half the world does not have clean water and millions are starving or malnourished. Very little of this reality is being shown through your media. For those wealthier nations, you are creating new "designer" foods in laboratories that are beginning to create health problems for yourselves and throughout nature, upsetting the natural flow and balance. Tampering with nature will not solve the problem of dwindling resources. It only serves to poison yourselves and the environment.

Your biggest challenge for many years to come will be population control. Wars and disease have always managed you in the past, but your numbers are exploding exponentially. A greater desire for peace along with modern medical advances have helped to increase your numbers. You have some very grim decisions looming on the horizon, but your numbers can be slowed down with proper planning. This does not mean a nuclear holocaust or creating new versions of the black plague.

You spay and neuter your pets to control their populations and prevent suffering, yet avoid the topic for yourselves when immense suffering is on the horizon. No nation, religion, or any group that needs numbers to remain strong is interested in being the first to step forward. You are in a mad dash toward self-destruction, with no restraint to stop you. Each group may see themselves as having the God-given right to reproduce, but if all are allowed to flourish, then your resources will be irreparably depleted in a few short generations. You are on your own with this problem; yet it is crucial that you solve it. The most important thing I can do is point out its urgency – the rest is up to you.

Culling the herd through violence will breed only more violence. Hatred and suspicion fuels your minds with the urgency to act in unloving ways. But if you start with love and compassion and act in preventative ways to curb your population rather than violent ones, things can be accomplished with more lasting success.

This idea begs the question as to who among you decides which people get "fixed" and who do not. You may do it by lottery, or it may be voluntary in exchange for some kind of benefit, but eventually it must be done. A world body must first be present to enforce it, however. And with the way your nations combat each other over much lesser issues, you will have your hands full. I am merely mapping out the concerns and future playing field that you absolutely must engage in to insure the survival of your species. No other species in the entire multi-billion year history of your world has ever been on such a hyper-speed crash course toward self-destruction than you currently are. The quickness involved with your downward spiral leaves little room for any higher intelligence examining the situation to conclude anything other than the fact that you are a suicidal species, or the most ignorant of the natural world and your place in it, than any other that has ever come before you. They would be justified to conclude that whatever sad outcome occurs would be exactly what you deserve. Enough of you are waking up, however, to the point where you may deserve to escape your leap into the abyss of ignorance. You are capable of great wisdom and are showing this wondrous aptitude – and not a moment too late. It is time for your awakening. You have been asleep for too long, and have amazing potential.

Exercise

Change the World
You cannot change the world until you first change yourself. By this time, as you have worked through the book, you should have experienced a change – anything from a slight shift to a major transformation. Something should have happened with you, whether you are conscious of it or not. Therefore, this exercise was saved for last. It is for those who have experienced a shift with this work and begun to make positive change. Now it is time to help change the world.

You may need someone to read this to you *slowly* as you go, much like a guided meditation; or you or this other person should read it slowly into a tape recorder and play it back whenever you need to (recommended). For the moment, if by yourself, read it through at least twice to remember the sequence you must follow. Then proceed.

Sit in a comfortable place and close your eyes. Settle your thoughts. Get relaxed. Take three deep breaths – in through the nose and out through the mouth. Slow your mind down. Stop it from jumping from one random thought to the next by focusing on your breath. Feel the air go through your nostrils. Follow it all the way into your lungs. Heighten your focus on the feeling of this air. For the first time, notice the difference in temperature as the air moves through your nose, throughout your passageway and into your lungs. Notice how cool it is compared to your body temperature. Hold each breath in your lungs for just a few extra seconds before expelling it through your mouth. Focus on the life-giving power of the oxygen, because without it, even for a couple of minutes, you would die. Become one with the air as it sits in your lungs. Then release it. As you expel it, notice that it has become the same temperature as your body. Hold your hand in front of your mouth and feel the warmth of your breath. It not only has your warmth, it also has your energy. From now on, whenever you say something important, invest the words with this spiritual energy, your life energy, by holding your breath inward for an extra second or two. Try this now, using a basic idea that you hold as important, like what you do as a career or something important that you plan to do. *Focus on your intent* before you speak if your words reflect something that you plan to do. Then release it with power, through your words, toward those who are listening – or out into the world. When the energy comes from deep within, it will have a far better impact when it reaches the outer world. The more you do this, the better at it you will become.

The exercise is over, but because it involves a different way of thinking and breathing, it must be practiced continually. Mankind is short on both patience and dedication, but if you exercise both traits with this exercise and in doing this course, you will benefit greatly. Many of you will fall by the wayside, but those who carry on with this exercise continually, for a few months, may find themselves in important, world-changing positions.

Even more possible will be your ability to effect positive change among your friends, family and within your communities. Doing so is of major importance, because the results will reverberate through the world in a type of snowball effect. A chain reaction of events often gets started in this way, with the results never being fully known by those who originally put things into motion. So if you think that you are just one person who does not make a difference in the big sheme of things, think again. Everything and everyone is connected, and changing the world does not mean that your ego has to be there, waiting to snatch the credit and experience instantaneous results. Part of your lesson on Earth is to learn how to work together for the greater good, rather than for personal gain. Use what you have learned in this course to put forth holistic ideas that do not require you to receive awards or personal recognition. If you receive them, that is fine – but it should never be your primary motivation.

This lesson has put you in contact with your soul. What you do with it, however, is up to you.

It takes a great soul to change the world and now that you have found it, you must *use* it to create change. You can no longer ignore it. Most people believe the soul resides somewhere in their heads because that is where they operate from, within their chattering minds. They "live" there. The mind is a soulless place because it is the home of the ego. Although the mind can sense the soul and do great things by connecting to it, it is not the home of the soul. The soul resides near the lungs and it can *charge your words*. Some of you believe that the soul resides in your heart. They are quite close in this assessment because your soul and the heart both work with the same electrical charge.

Your heart is an electrical instrument and has an electrical charge. That is why when someone has heart failure in a hospital emergency room, an electrical charge is administered to start it back up again. If you are swimming in the ocean, creatures like the great white shark can detect the electrical charge in each beat of your heart. But you are aware of nothing and detect nothing, unless you use advanced instrumentation.

You are clueless about your souls and its power. A secret that may help your scientists is that this *measurable* electrical energy within the body has a connection to the soul. This will lead you to one day "find" the soul after centuries of searching, although it will be accomplished using "non-conceptual scientific means." This term may seem contradictory, but you will one day understand.

Your life energy, sometimes called kundalini, rests in most of you at the base of the spine. When it gets used it never usually goes far, being expended in the sexual process through the lower energy centers, or chakras. It is only natural that your life energy be used, in its basest form, in creating more life. However, the kundalini can sometimes be raised up in people to the point of the heart chakra, where the soul resides, and even higher. When it reaches up to the brow chakra (often called the "third eye" in mystic circles for many centuries), between and behind your eyes, then and only then does the *realization* of the soul and its power occur to you. This is the first place where the soul energy links to the brain, often flooding it with light and giving you the realization that was, until this time, completely hidden. An awakening often occurs with the setting off of this electrical "spark" of sorts into the circuitry of the brain. It fools people into believing that the mind is the home of the soul, but it is the mind that merely experiences its' realization (not its' location). Your soul is not entirely electrical, but there is a connecting "circuit" there, at this inner third eye point, between the mind and the soul.

The exercise just covered reveals exactly how to open this circuit and use it at a lower level, without necessarily raising the kundalini to the third eye, which you may not be ready for. That happens only when the mind and body are ready, but in the meantime you can use this similar electrical charge by itself, with your words, if you know how to do it.

Your messages sent via the media are transmitted through electricity, whether it be through TV, the radio, telephones, ipods or the Internet. When you charge the air in your lungs with your intent, it will charge the words you use and help to *create* the results of that intent when people hear and experience them as they do through the media. You of course must follow your words with *determined action*, but people will tend to support you because they respond far better to words charged with energy. The soul has an electrical component to it. Therefore, when a person has finished an inspiring talk some will say, "That speech was electric!" because they could feel the energy in the room.

Very few know how to create this energy. You will if you practice. Use it wisely. Use it only for good and never try to harm others in any way. Your intent should never be negative or harmful. If someone intuitively feels your energy is not right, then it will get blocked or deflected and you will have no effect. And if you know someone is up to no good and using this exercise, you can deflect the bad energy back to them, to their detriment, after doubling it. You just focus it back toward them with intense conscious effort.

The soul is inherently good. Only the mind can make the soul negative. My *intent* is for you to know how to create the positive through direct contact with the soul, not allowing your mind to get in the way with anything negative. What I have revealed should not be used to harm others in any way. This then, will result in your true Self coming through. When you charge that energy and transmit it to others, it will be experienced as "electric." People will listen. Many will be inspired and act upon your words. You will be changing the world through a direct alignment with your soul – and with those who resonate with it.

Summary

There is another dimension to humankind and within your very self that is unseen, and has remained shrouded over for centuries. You have been living in darkness, but are moving slowly toward the light. At times you have taken a wrong turn, away from the light. This is one of those times. My work as the Overseer can sometimes correct a wrong turn, and that has been the goal of this short and modest work.

Postscript

You have now completed the lessons. If you have done them with conviction then you can see how much you have grown. Some of the exercises at the end of the lessons are meant to be experienced on a continuous basis; otherwise they are not as useful. Be conscious of this fact. Continue with the exercises that serve you. One thing should be certain by now. You are no longer the same person that you were when you started the book. A door has opened and you have begun to walk through it.

This does not mean, however, that you walk away from this book. For a while, it should function as your guide. Mark the pages or your calendar with the useful exercises and keep doing them. Self-discipline is one of the hardest things for individual humans to cultivate. But if you can be successfully disciplined with these exercises, then the rewards may shock you. Do you ever wonder how a caterpillar feels to metamorphosize out of its cocoon and suddenly sprout wings? It is truly shocking and painful to become an entirely new and different creature – but they feel a tremendous *freedom* as well. Suddenly, they can fly. Your transformation will not be as physically evident, but it is just as real. As you continue with the exercises, there will come a time when you will no longer need the book as your cocoon. You will fly. You will then be *living* the lessons, naturally. And if you get to the point of living these lessons naturally, you will find yourself as a powerful force in the world. I will have accomplished my goal, by having helped you in this regard. And you will be on the right path for your life. There will be no doubts for you because the exercises, if done properly, will bring you in contact with the true force that guides you. This book is like the cocoon, preparing you and nurturing you. I want to see humanity fly, rather than fail or self-destruct.

Therefore, you are encouraged to start classes if you feel led to do so. Create groups and go through this book together. Compare notes, do the exercises together, plot your progress and experience your growth and power as a group. A synergy will develop, ideas will be exchanged and you will end up accomplishing greater things with the help of others. It is good to do this book alone, but amazing things can occur when others are on the path with you. Your creativity will be sparked within your group, you will inspire each other to greater heights and powerful new ideas will seem to come out of nowhere. They will send you in directions that will change your life, the lives of those in your group, as well as the lives of the others you touch.

This can still occur if you continue with the exercises alone, but developing self-discipline within a group is easier to do. Self-discipline has always been a major human challenge and is often the key to success. Whether you enter into a group or not, your journey is not over – it has only just begun. Make it count, make it matter; make a difference.

This book has offered you a new way to connect with the very deepest part of yourself. From there, it can allow you to discover your true identity and the proper path for you to follow, thereby giving you the *power* to make a positive difference in the world. For many of you, there will be no better way to do this than through the use of this book. It contains all of the tools you need, but you must keep using them. For example, it is not practical to build the foundation for a house, and then move in and declare that it is done. Continue with the exercises. They are intentionally few and easy, so that your "toolbox" will be light and you can continue building. Follow the same pattern you used the first time through, or adjust it for your comfort. It may help to use a calendar and the list of exercises on page 8 for reference. You have a magnificent structure that awaits completion by using short, powerful exercises that are deceptively simple and easy to do.

Who is the Overseer? It is not important. I look upon you from above, but you may now look upon yourselves from within. You may now be in touch with your own personal Overseer – if you have found your wisdom and are using it. That has been the purpose of this book. I look upon many, but each of you has your own Overseer or higher Self, that looks upon only you. It is not a fantasy, which you may have realized with my help.

With the lessons – or at least your first reading of them – being over, the *purpose for your life* has just begun. You must keep using your energy, wisdom and the exercises in this book to fulfill your mission in life. It will be your true mission, not what society and the economic powers have "herded" you into doing, like a sheep, for their own benefit. It is time for you to be you. Allow your Overseer to guide you by staying in alignment with your power.

You are all connected with each other and the more who connect together, consciously and with intent, the more you can accomplish. It starts with you and you now have access to that power. Use it, together, or your place on the planet will slip away. If you do not balance yourselves with nature, the Earth will claim you, as it has done with numerous vanished species for eons. You are only important to yourselves. Make yourselves important to the Earth, or she will take you.

The Overseer has finished delivering your lessons and challenges. He has completed the task. The rest is up to you.

WHAT OTHERS ARE SAYING

The Overseer has appeared at crucial times throughout history, when mankind has needed him. Sometimes they listen; sometimes they don't. Because of the times we live in, this book is the Overseer's most definitive statement ever. Everyone now has the chance to listen, learn and progress. The *Lessons* are so important that the following historical figures have stepped forward from the past to speak.

I ate forbidden fruit, considered to be the biggest mistake in history. However, it is not too late. The Overseer can correct this blunder and the mistakes of those who followed me. —Adam

If I had listened to the Overseer I would not have gotten so lost and confused. I thought I was in India. Instead of naming them Indians I could have left them alone, found the right place, and avoided the eventual slaughter and slavery of millions of innocent people.
—Chistopher Columbus

I listened to the Overseer, set many people free and changed history for the better. —Abraham Lincoln

I knew the Overseer. He was with me all along.
—Albert Einstein

Overseer? What Overseer?   —Adolf Hitler

ABOUT THE AUTHOR

The Overseer lives quietly in a higher dimension, but has been known to visit this one. He is immortal, but is not God. He continues to do what he has always done for thousands of years, which is to oversee the development of mankind within the framework of God's plan. His main job is to prevent humanity from acting on its self-destructive tendencies. His wife Kina oversees nature, so he must work hard to protect her from humanity's ignorance. They have four children: Earth, Air, Fire and Water, and a blue-eyed sabre-toothed cat named Caesar, who roams the starry skies. This is the Overseer's first book – written with lightning bolts from his extended finger in about one eight of a second. He can be contacted by mail through the publisher, and will respond through vivid dreams.

ALSO AVAILABLE FROM THE BOOK TREE

NEXUS Magazine offers you all of the information that the mainstream media will not tell you. It is a fascinating magazine out of Australia with worldwide distribution, so check your local newsstand or call the number below.